Introduction to the student

This homework book is designed to give you extra practice on the topics covered in class. Each exercise in *Higher Mathematics for GCSE* (second edition) has a corresponding homework exercise of the same format in this book.

No formulae or worked examples are given in this book. We assume that these have been provided by your teacher and if you have worked from the Higher book in class, you should have sufficient reference material in your own notes.

Most exercises include an 'examination-type question' and these are highlighted with an asterisk ★ in the margin. These questions give you the opportunity to see how each topic might appear in your examinations.

We wish you every success!

Keith Gordon, Brian Speed and Kevin Evans
July 2001

Mathematics for GCSE (second edition) text books and homework resources have been written by senior examiners with many years of teaching experience to provide comprehensive preparation for GCSE Mathematics courses starting from 2001. Though written specifically for AQA specifications A and B, the books are also suitable for the new Edexcel and OCR qualifications.

Published by Collins Educational
77-85 Fulham Palace Road
Hammersmith
London W6 8JB

Browse the complete Collins catalogue at
www.collinseducation.com

Reprinted 2001, 2002, (twice), 2003, 2004, 2005

10 9 8 7

ISBN 0 00 712364 7

British Cataloguing in Publication Data
A catalogue record for this book is available from the British Library

Edited by Simon Gerratt
Typesetting by Derek Lee
Illustrations by Moondisks, Cambridge; Illustrated Arts, Sutton; Barking Dog Art; Mark Jordan and Simon Gerratt
Cover by Sylvia Kwan, Chi Leung
Production by Kathryn Botterill
Commissioned by Mark Jordan
Printed and bound by Martins the Printers, Berwick upon Tweed

Homework 1A

1 Increase each of the following by the given amount.
   a £80 by 5% b 14 kg by 6% c £42 by 3%

2 Increase each of the following by the given amount.
   a 340 g by 10% b 64 m by 5% c £41 by 20%

3 Keith, who was on a salary of £34 200, was given a pay rise of 4%. What was his new salary?

4 In 1991 the population of Dripfield was 14 200. By 2001 that had increased by 8%. What was the population of Dripfield in 2001?

5 In 1993 the number of bikes on the roads of Doncaster was about 840. Since then it has increased by 8%. Approximately how many bikes are on the roads of Doncaster now?

6 At a nightclub there are always about 30% more girls than boys. If at one disco there were 40 boys, how many girls were there?

UAM ★7 A house was priced at £42 000 in January 1999. In January 2000, it was 4% more expensive. In January 2001, it was 5% more expensive than the price in 2000. What was the price of the house in January 2001?

Homework 1B

1 Decrease each of the following by the given amount.
   a £8 by 4% b 17 kg by 6% d 240 m by 2%

2 Decrease each of the following by the given amount.
   a 360 g by 10% b 440 m by 15% c 360 cm by 25%

3 A van valued at £8 400 last year is now worth 12% less. What is its value now?

4 A firm employed 80 workers. But it had to streamline its workers and lose 20% of the workers. How many workers does the firm have now?

5 On the first day of a new term, a school expects to have an attendance rate of 99%. If the school population is 700 pupils, how many pupils will the school expect to be absent on the first day of the new term?

★6 Most weighing scales in the home have an error of about 10% from the true reading. When my weighing scales say 500 grams, what is the
   a lowest weight it could represent b largest weight it could represent?

7 By putting cavity wall insulation into your home, you could use 20% less fuel. A family using an average of 850 units of electricity a year put cavity wall insulation into their home. How much electricity would they expect to use now?

PROOF 8 Prove that a 10% increase followed by a 10% decrease is equivalent to a 1% decrease overall.

HOMEWORK 1C

1 A small plant increases its height by 10% each day for the second week of its growth. At the end of the first week, the plant was 5 cm high.
What is its height after a further
a 1 day **b** 2 days **c** 4 days **d** 1 week?

2 The headmaster of a new school offered his staff an annual pay increase of 5% for every year they stayed with the school.
a Mr Speed started teaching at the school on a salary of £28 000. What salary will he be on after 3 years if he stays at the school?
b Miss Tuck started teaching at the school on a salary of £14 500. How many years will it be until she is earning a salary of over £20 000?

3 Billy put a gift of £250 into a special savings account that offered him 8% compound interest if he promised to keep the money in for at least 2 years. How much was in this account after
a 2 years **b** 3 years **c** 5 years?

4 The penguin population of a small island was only 1500 in 1998, but it steadily increased by about 15% each year. Calculate the population in
a 1999 **b** 2000 **c** 2002

★5 A sycamore tree is 40 cm tall, it grows at a rate of 8% per year. A conifer is 20 cm tall. It grows at a rate of 15% per year. How many years does it take before the conifer is taller than the sycamore?

HOMEWORK 1D

1 Express the following as percentages.
a £4 of £16 **b** £3 of £50 **c** 25 kg of 500 kg
d 12 hours of 1 day **e** 15 minutes of 1 hour **f** 13 m of 20 m

2 Find the percentage profit on the following.

	Item	*Retail price (selling price)*	*Wholesale price (price the shop paid)*
a	DVD	£210	£150
b	Widescreen TV	£840	£500

3 There were 4 left-handed pupils in a class of 25. What percentage were left-handed?

4 James came home from school one day with his end-of-year test results. Change each of John's results to a percentage.

Maths	63 out of 75	English	56 out of 80
Science	75 out of 120	French	27 out of 60

★5 Kris had an annual salary of £32 000 in 2000, which was increased to £33 600 in 2001. What percentage increase does this represent?

6 During the wet year of 1993, it rained in London on 80 days of the year. What percentage of days were wet? Give your answer to two significant figures.

HOMEWORK 1E

1 Find what 100% represents when
a 20% represents 160 g **b** 25% represents 24 m **c** 5% represents 42 cm

2 Find what 100% represents when
a 40% represents 28 kg **b** 30% represents £54 **c** 15% represents 6 hours

3 VAT is a government tax added to goods and services. With VAT at 17.5%, what is the pre-VAT price of the following priced goods?
Jumper £14.10 Socks £1.88 Trousers £23.50

4 Paula spends £9 each week on CDs. This is 60% of her weekly income. How much is Paula's weekly income?

5 Alan's weekly pay is increased by 4% to £187.20. What was Alan's pay before the increase?

6 Kev sold his bike for £60, making a profit of 20% on the price he paid for it. How much did Kev pay for the bike?

★7 Gran used 240 g of mixed fruit and nut in a cake. This represented 30% of the weight of the cake. How much did the cake weigh?

8 99 Rock music CDs represents just 18% of my entire CD collection. How many CDs have I?

HOMEWORK 1F

Examples. 5.852 will round off to 5.85 to two decimal places
7.156 will round off to 7.16 to two decimal places
0.284 will round off to 0.3 to one decimal place
15.3518 will round off to 15.4 to one decimal place

1 Round off each of the following numbers to one decimal place.
a 3.73 **b** 8.69 **c** 5.34 **d** 18.75 **e** 0.423
f 26.288 **g** 3.755 **h** 10.056 **i** 11.08 **j** 12.041

2 Round off each of the following numbers to two decimal places.
a 6.721 **b** 4.457 **c** 1.972 **d** 3.485 **e** 5.807
f 2.564 **g** 21.799 **h** 12.985 **i** 2.302 **j** 5.555

3 Round off each of the following to the number of decimal places indicated.
a 0.085 (2 dp) **b** 4.558 (2 dp) **c** 2.099 (2 dp) **d** 0.7629 (3 dp)
e 7.124 (1 dp) **f** 8.903 (2 dp) **g** 23.7809 (3 dp) **h** 0.99 (1 dp)

4 Round off each of the following to the nearest whole number.
a 6.7 **b** 9.3 **c** 2.8 **d** 7.5 **e** 8.38
f 2.82 **g** 2.18 **h** 1.55 **i** 5.252 **j** 3.999

HOMEWORK 1G

1 Round off each of the following numbers to 1 significant figure.

a	46 313	**b**	57 123	**c**	30 569	**d**	94 558	**e**	85 299
f	54.26	**g**	85.18	**h**	27.09	**i**	96.432	**j**	167.77
k	0.5388	**l**	0.2823	**m**	0.005 84	**n**	0.047 85	**o**	0.000 876
p	9.9	**q**	89.5	**r**	90.78	**s**	199	**t**	999.99

2 What is the least and the greatest number of people that can be found in these towns?

Hellaby	population 900 (to 1 significant figure)
Hook	population 650 (to 2 significant figures)
Hundleton	population 1050 (to 3 significant figures)

3 Round off each of the following numbers to 2 significant figures.

a	6725	**b**	35 724	**c**	68 522	**d**	41 689	**e**	27 308
f	6973	**g**	2174	**h**	958	**i**	439	**j**	327.6

4 Round off each of the following to the number of significant figures (sf) indicated.

a	46 302 (1 sf)	**b**	6177 (2 sf)	**c**	89.67 (3 sf)	**d**	216.7 (2 sf)
e	7.78 (1 sf)	**f**	1.087 (2 sf)	**g**	729.9 (3 sf)	**h**	5821 (1 sf)
i	66.51 (2 sf)	**j**	5.986 (1 sf)	**k**	7.552 (1 sf)	**l**	9.7454 (3 sf)
m	25.76 (2 sf)	**n**	28.53 (1 sf)	**o**	869.89 (3 sf)	**p**	35.88 (1 sf)
q	0.084 71 (2 sf)	**r**	0.0099 (2 sf)	**s**	0.0809 (1 sf)	**t**	0.061 97 (3 sf)

HOMEWORK 1H

1 Find approximate answers to the following sums.

a	4324×6.71	**b**	6170×7.311	**c**	72.35×3.142
d	4709×3.81	**e**	$63.1 \times 4.18 \times 8.32$	**f**	$320 \times 6.95 \times 0.98$
g	$454 \div 89.3$	**h**	$26.8 \div 2.97$	**i**	$4964 \div 7.23$
j	$316 \div 3.87$	**k**	$2489 \div 48.58$	**l**	$63.94 \div 8.302$

2 Find the approximate monthly pay of the following people whose annual salary is

a Joy £47 200 **b** Amy £24 200 **c** Tom £19 135

3 Find the approximate annual pay of these brothers who earn:

a Trevor £570 a week **b** Brian £2728 a month

4 A litre of creosote will cover an area of about 6.8 m^2. Approximately how many litre cans will I need to buy to creosote a fence with a total surface area of 43 m^2?

★5 A groundsman bought 350 kg of seed at a cost of £3.84 per kg. Find the approximate total cost of this seed.

6 A greengrocer sells a box of 250 apples for £47. Approximately how much did each apple sell for?

7 Keith runs about 15 km every day. Approximately how far does he run in

a a week **b** a month **c** a year?

HOMEWORK 1I

1 Round off each of the following figures to a suitable degree of accuracy.
  - a Kris is 1.6248 metres tall.
  - b It took me 17 minutes 48.78 seconds to cook the dinner.
  - c My rabbit weighs 2.867 kg.
  - d The temperature at the bottom of the ocean is 1.239 °C.
  - e There were 23 736 people at the game yesterday.

2 How many jars each holding 119 cm^3 of water can be filled from a 3 litre flask?

3 If I walk at an average speed of 62 metres per minute, how long will it take me to walk a distance of 4 km?

4 Helen earns £31 500 a year. How much does she earn in
  - a 1 month b 1 week c 1 day?

5 Dave travelled a distance of 350 miles in 5 hours 40 minutes. What was his average speed?

6 Ten grams of Gold cost £2.17. How much will one kilogram of Gold cost?

Chapter 2 Shape

HOMEWORK 2A

1 Find the circumference of each of the following circles, round off your answers to 1 dp.
  - a Diameter 3 cm b Radius 5 cm c Radius 8 m
  - d Diameter 14 cm e Diameter 6.4 cm f Radius 3.5 cm

2 John runs twice round a circular track which has a radius of 50 m. How far has he run? Give your answers in terms of π.

3 A rolling pin has a diameter of 5 cm.
  - a What is the circumference of the rolling pin?
  - b How many revolutions does it make when rolling a length of 30 cm?

4 What is the total perimeter of a semicircle of diameter 7 cm? Give your answer to 1 dp.

5 What is the total perimeter of a semicircle of radius 6 cm? Give your answer in terms of π.

6 How many complete revolutions will a bicycle wheel, radius 28 cm, make in a journey of 3 km?

★7 A circle has a circumference of 12 cm. What is its diameter?

HOMEWORK 2B

1 Calculate the area of each of these circles, giving your answers to 1 decimal place, except for **a** and **d**, where your answer should be in terms of π.
  - a Radius 4 cm b Diameter 14 cm c Radius 9 cm
  - d Diameter 2 m e Radius 21 cm f Diameter 0.9 cm

2 A garden has a circular lawn of diameter 20 m. There is a path 1 m wide all the way round the circumference. What is the area of this path?

3 Calculate the area of a semi-circle with a diameter of 15 cm. Give your answer to 1 dp.

4 A circle has an area of 50 m^2. What is its radius?

★5 I have a circle with a circumference of 25 cm. What is the area of this circle? Give your answer to 1 decimal place.

6 Jane walked around a circular lawn. She counted 153 paces to walk round it. Each of her paces was about 42 cm. What is the area of the lawn?

Homework 2C

1 For these sectors, calculate the arc length and the sector area.

a

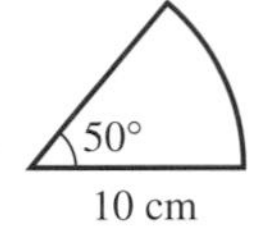

b

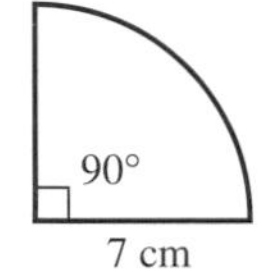

2 Calculate the arc length and the area of a sector whose arc subtends a right angle of a circle of diameter 10 cm. Give your answer in terms of π.

3 Calculate the total perimeter of each of these shapes.

a

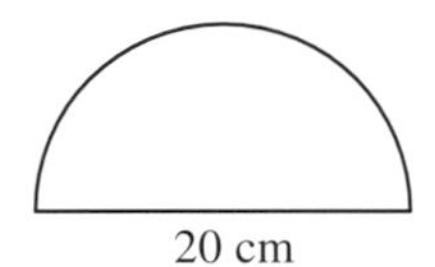

b

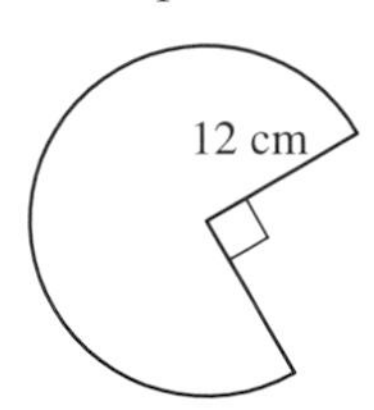

4 Calculate the area of each of these shapes.

a

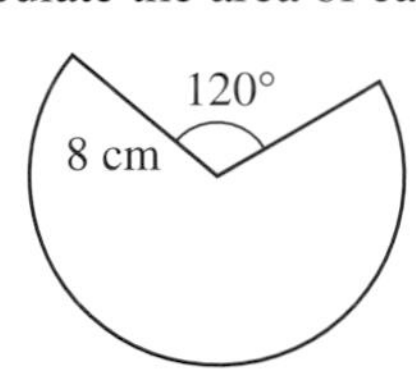

b

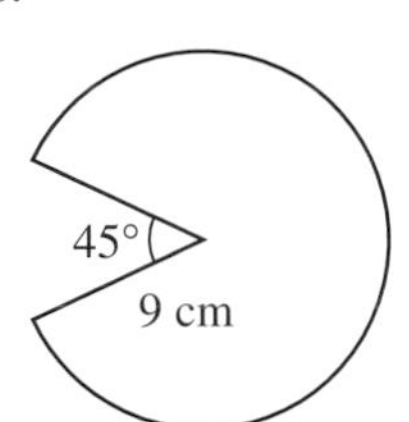

★5 There is an infra red sensor in a security system. The sensor can detect movement inside a sector of a circle. The radius of the circle is 16 m. The sector is 120°. Calculate the area of the sector.

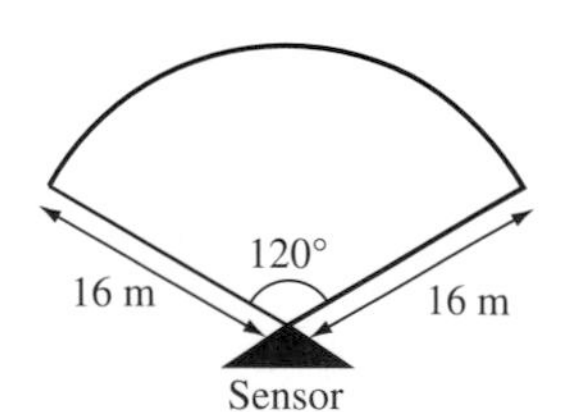

Homework 2D

1 Calculate the perimeter and the area of each of these trapeziums.

a

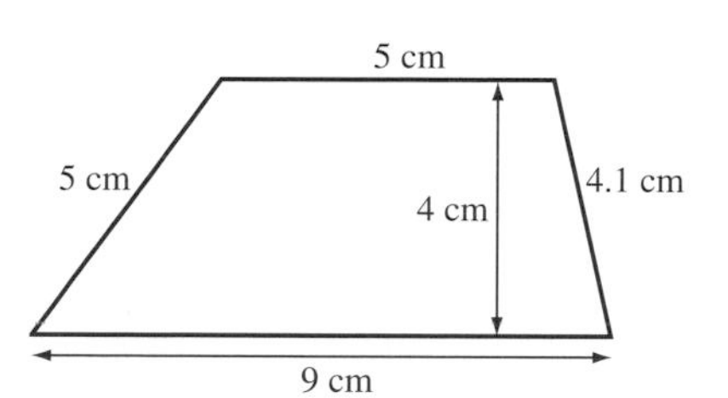

b

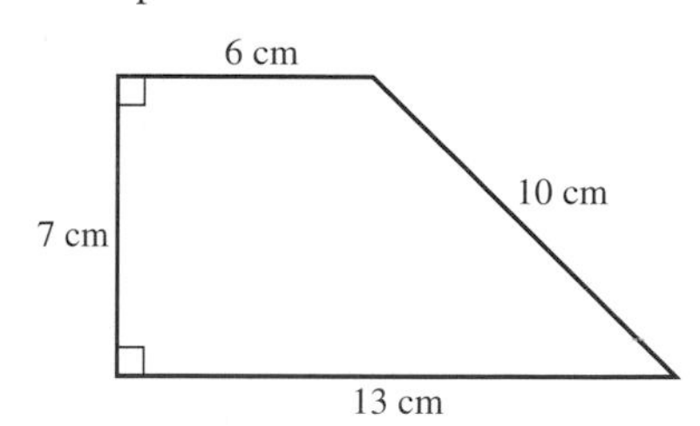

2 Calculate the area of each of these shapes.

a

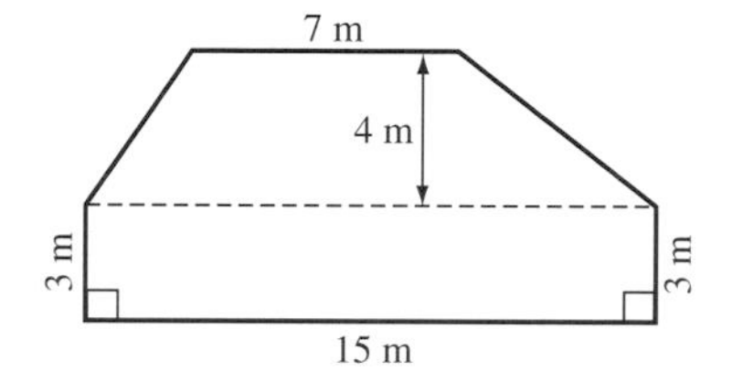

b

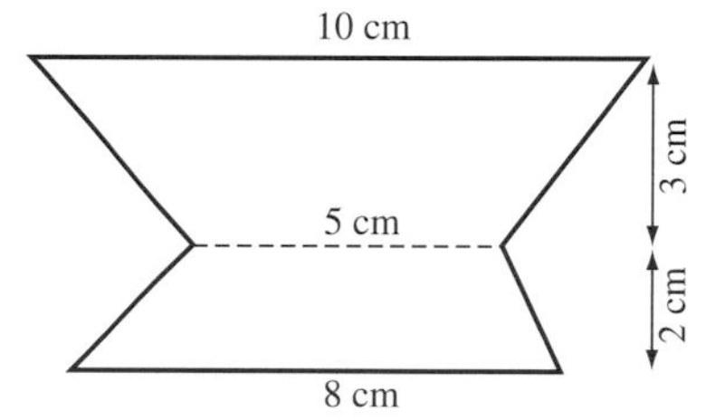

3 Calculate the area of the shaded part in each of these diagrams.

a

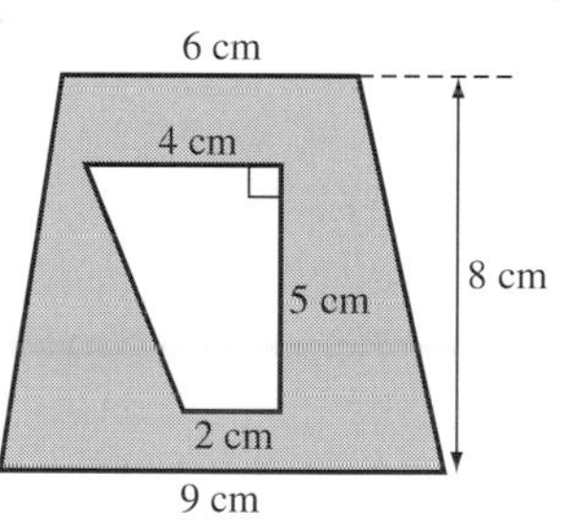

b

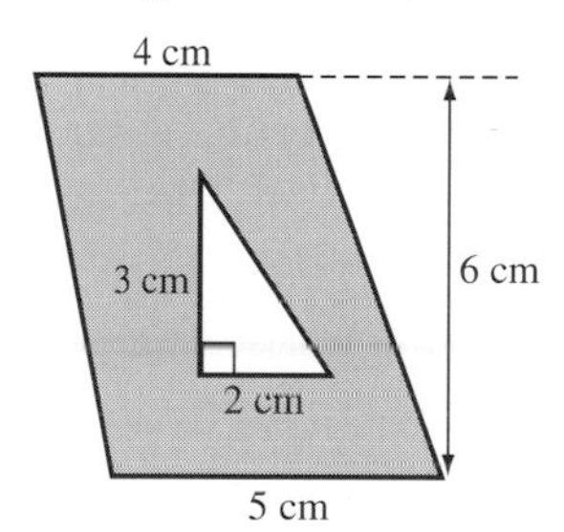

4 Which of the following shapes has the largest area?

a

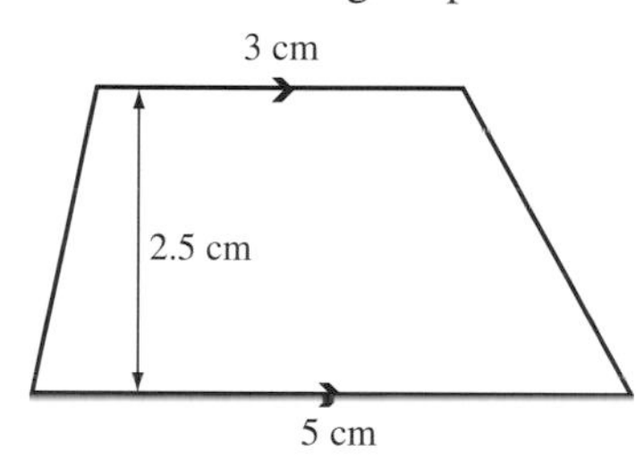

b

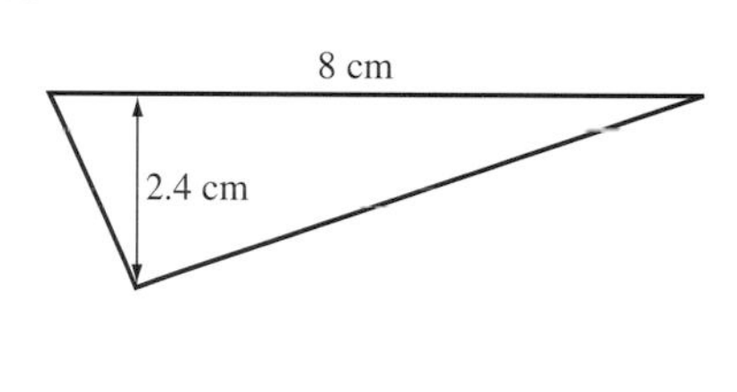

5 What percentage of this shape has been shaded?

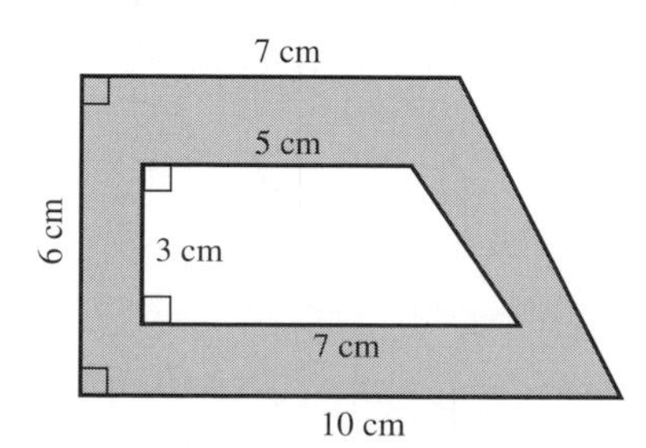

HOMEWORK 2E

1 Find the density of a piece of wood weighing 135 g and having a volume of 150 cm^3.

2 Calculate the density of a metal if 40 cm^3 of it weighs 2500 g.

3 Calculate the weight of a piece of plastic, 25 cm^3 in volume, if its density is 1.2 g/cm^3.

4 Calculate the volume of a piece of wood which weighs 350 g and has a density of 0.9 g/cm^3.

5 Find the weight of a marble statue, 540 cm^3 in volume, if the density of marble is 2.5 g/cm^3.

6 Calculate the volume of a liquid weighing 1 kg and having a density of 1.1 g/cm^3.

7 Find the density of the material of a stone which weighs 63 g and has a volume of 12 cm^3.

8 It is estimated that a huge rock balanced in the ceiling of a cave has a volume of 120 m^3. The density of the rock is 8.3 g/cm^3. What is the estimated weight of the rock?

9 A 1 kg bag of flour has a volume of about 900 cm^3. What is the density of flour in g/cm^3?

HOMEWORK 2F

1 Find **i** the volume and **ii** the curved surface area of a cylinder with base radius 5 cm and height 4 cm. Give your answer in terms of π.

2 Find **i** the volume and **ii** the curved surface area of a cylinder with base radius 8 cm and height 17 cm. Give your answer to a suitable degree of accuracy.

3 Find **i** the volume and **ii** the total surface area of each of these cylinders.

a

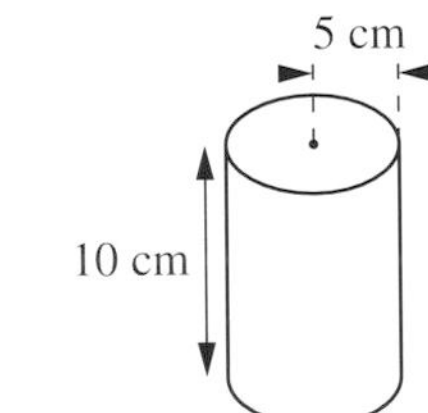

b

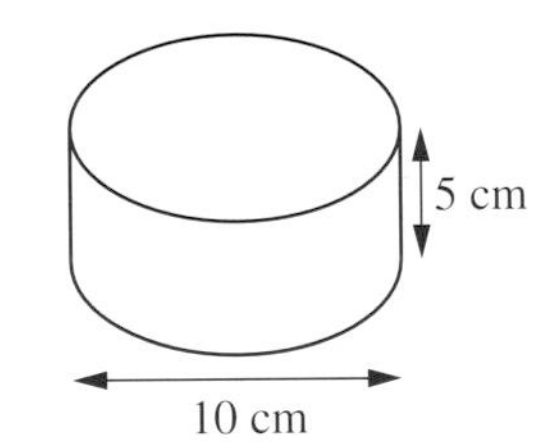

4 What is the radius of a cylinder, height 6 cm, with a volume of 24π cm^3?

5 What is the radius of a cylinder, height 10 cm, with a curved surface area of 360π cm^2?

6 What is the height of a cylinder, diameter 12 cm, with a volume of 108π cm^3?

7 A cylinder of height 20 cm has a curved surface area of 200 cm^2. Calculate the volume of this cylinder.

8 Calculate the curved surface area of a cylinder which has a height of 18 cm and a volume of 390 cm^3.

★9 A cylinder has the same height and radius. The total surface area is 100π. Calculate the volume. Give your answer in terms of π.

Homework 2G

1 For each prism shown, calculate the area of the cross-section and the volume.

a

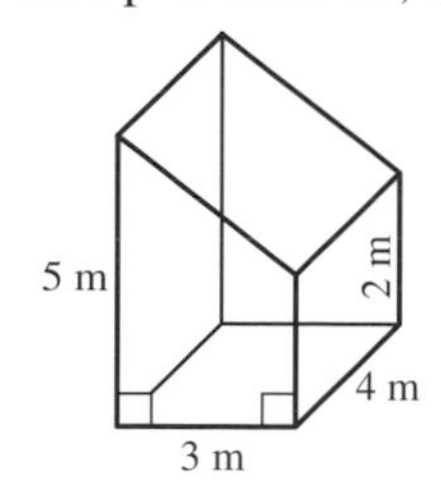

b

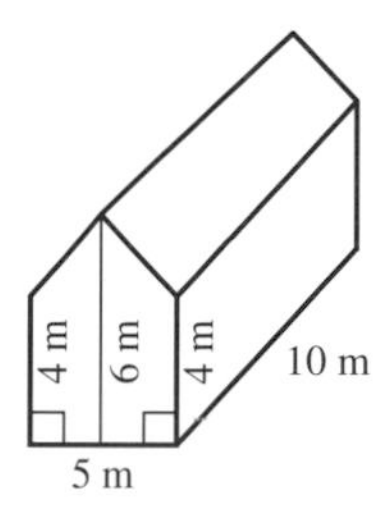

2 Which of these solids is

a the heaviest

b the lightest?

i

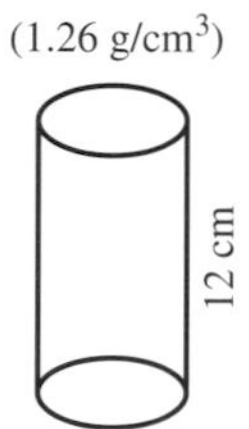

ii

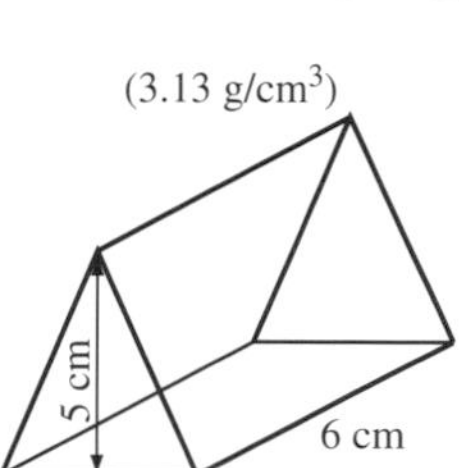

iii

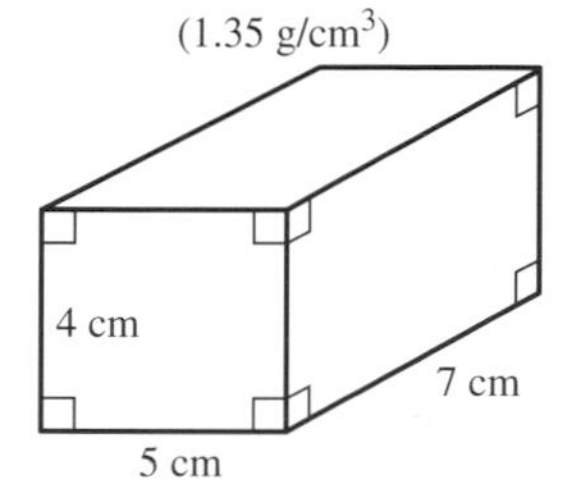

Homework 2H

1 Calculate the volume of each of these pyramids, all with rectangular bases.

a

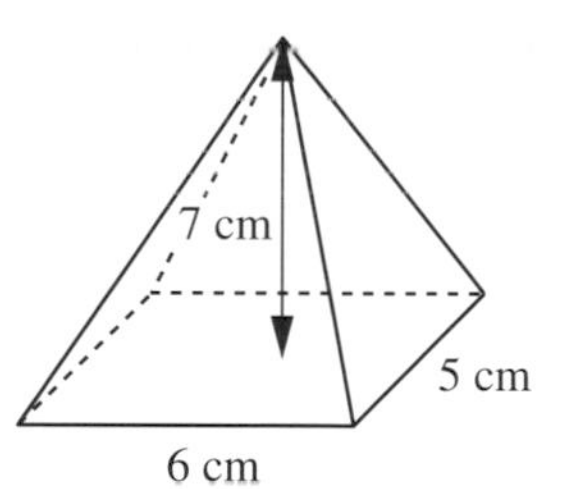

b

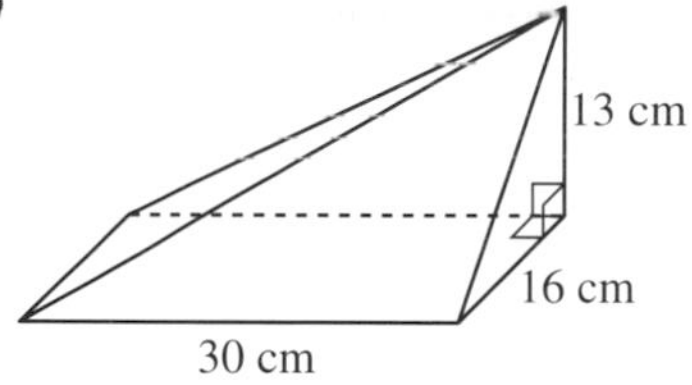

2 Calculate the volume of a pyramid having a square base of side 10 cm and a vertical height of 18 cm.

3 Calculate the volume of this shape.

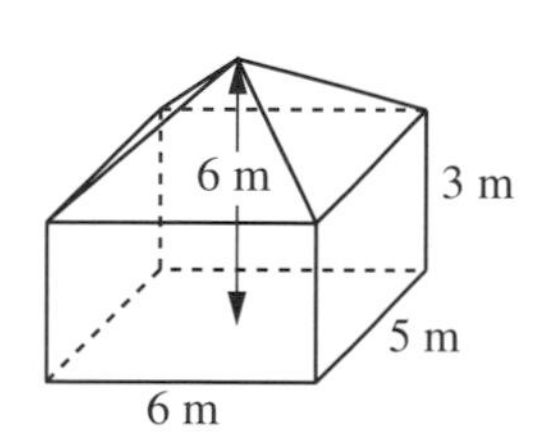

4 Calculate the height h of a rectangular-based pyramid with a length of 14 cm, a width of 10 cm and a volume of 140 cm^3.

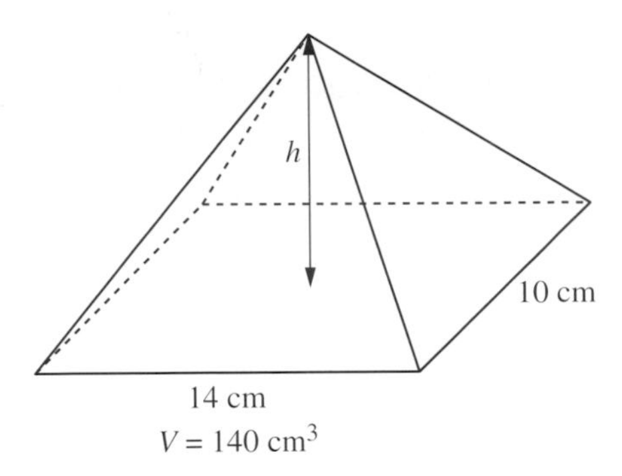

★5 The pyramid in the diagram has its top half cut off as shown. The shape which is left is called a frustum. Calculate the volume of the frustum.

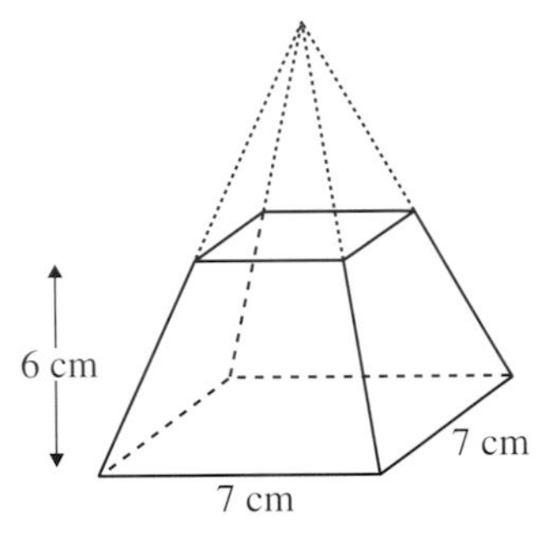

HOMEWORK 2I

1 For each cone, calculate **i** its volume and **ii** its total surface area. (The units are cm.)

a

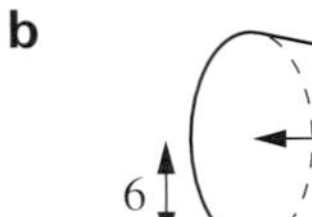

b

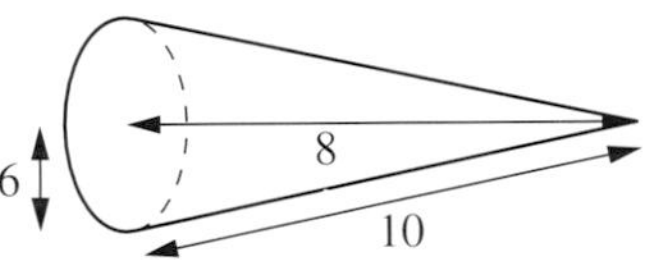

2 Find the total surface area of a cone whose base radius is 4 cm and slant height is 6 cm. Give your answer in terms of π.

3 Find the volume of a cone whose base radius is 6 cm and vertical height is 8 cm. Give your answer in terms of π.

4 In order to make a cone, a sector of angle 60° is cut from a circle whose radius is 12 cm.

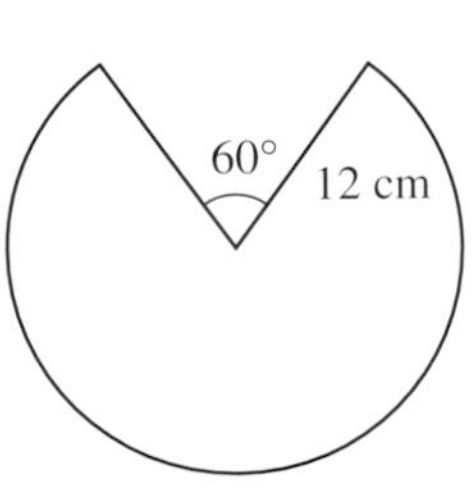

a Calculate the circumference of the base of the cone.
b Calculate the radius of the base of the cone.
c State the length of the slant height of the cone.
d Calculate the curved surface area of the cone.
e Calculate the vertical height of the cone.
f Calculate the volume of the cone.

5 A container in the shape of a cone, base radius 10 cm and vertical height 19 cm, is full of water. The water is poured into an empty cylinder of radius 15 cm. How high is the water in the cylinder?

★6 The diagram shows a paper cone. The diameter of the base is 4.8 cm and the slant height is 4 cm. The cone is cut along the line AV and opened out flat, as shown below.

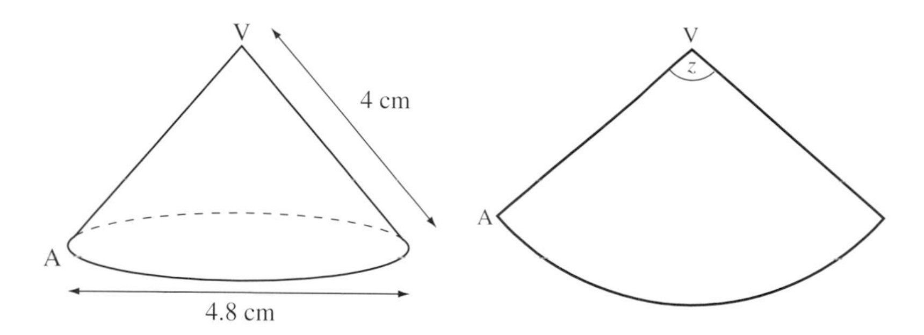

Calculate the size of angle z.

HOMEWORK 2J

1 Calculate the volume of each of these spheres. Give your answers in terms of π.

a Radius 3 cm **b** Diameter 30 cm

2 Calculate the surface area of each of these spheres. Give your answers in terms of π.

a Radius 4 cm **b** Diameter 10 cm

3 Calculate the volume and the surface area of a sphere with a diameter of 30 cm. Give your answers to a suitable degree of accuracy

4 Calculate, correct to 1 decimal place, the radius of a sphere

a whose surface area is 200 cm^2 **b** whose volume is 200 cm^3.

5 The volume of a sphere is 50 m^3. Find its diameter.

6 What is the volume of a sphere whose surface area is 400 cm^2?

★7 A spinning top which consists of a cone of base radius 6 cm, height 10 cm and a hemisphere of radius 6 cm, is illustrated on the right. Give your answers in terms of π.

a Calculate the volume of the spinning top.

b Calculate the total surface area of the spinning top.

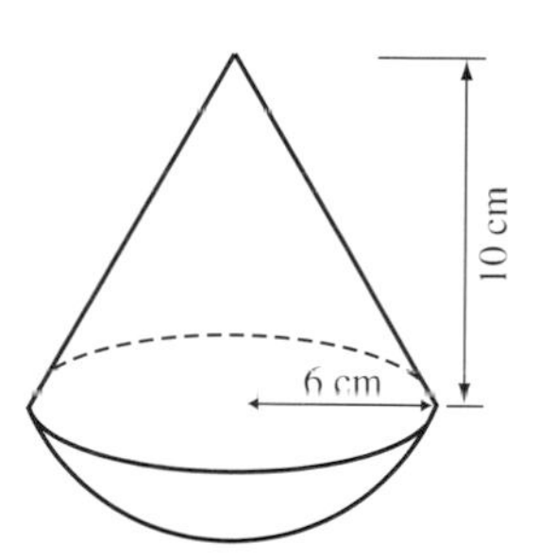

Chapter 3 Algebra 1

HOMEWORK 3A

1 Write down the algebraic expression that says:

a 3 more than x **b** 7 multiplied by k **c** 4 less than x
d k multiplied by t **e** k more than y **f** x divided by 3
g t less than y **h** 5 divided by x **i** h multiplied by itself

2 Write these expressions in a shorter form.

a $a + a + a + a$ **b** $b + b + b + b + b$ **c** $c + c + c + c + c + c + c$
d $2d + 5d$ **e** $6e + 5e$ **f** $f + 2f + 3f + 4f$
g $4v - 6v$ **h** $4x + 5x - 6x + x$ **i** $3z - 4z + 2z - z$

3 Simplify the following expressions

a $5y + 3x + 6y + 4x$ **b** $2m + p - m + 5p$ **c** $4x + 7 + 3x - 5$
d $9 - 3x - 2 + 6x$ **e** $5p + 3t + p - t$ **f** $7 + x + 5x - 3$

4 Multiply out the following brackets, leaving the answer as simple as possible.

a $5(f + 2)$ **b** $4(k - 1)$ **c** $3(t + 2)$ **d** $8(3d + 2)$
e $3(2t - 5)$ **f** $7(2m + 5)$ **g** $4(3 + 5w)$ **h** $5(2 - 3x)$

5 Expand and simplify each of the following.

a $3(2x + y) + 4(x + 3y)$ **b** $6(3t + 2p) + 3(4t + 3p)$ **c** $3(4t - 5q) + 6(t - 2q)$
d $2(4p + t) + 4(2p + 3t)$ **e** $5(2t - 3n) - 3(5t + n)$ **f** $3(2p + 4t) - 4(p - 5t)$

6 A girl went shopping with £P. She spent £T. How much has she got left?

7 4 beakers cost £B. What is the cost of one beaker?

HOMEWORK 3B

1 Find the value of $4x + 3$ when **i** $x = 3$ **ii** $x = 6$ **iii** $x = 11$

2 Find the value of $3k - 1$ when **i** $k = 2$ **ii** $k = 5$ **iii** $k = 10$

3 Find the value of $4 + t$ when **i** $t = 5$ **ii** $t = 8$ **iii** $t = 15$

4 Evaluate $14 - 3f$ when **i** $f = 4$ **ii** $f = 6$ **iii** $f = 10$

5 Evaluate $6m + 5$ when **i** $m = 1$ **ii** $m = 7$ **iii** $m = 11$

6 Evaluate $\frac{4d - 7}{2}$ when **i** $d = 2$ **ii** $d = 5$ **iii** $d = 15$

7 Find the value of $5x + 2$ when **i** $x = -2$ **ii** $x = -1$ **iii** $x = 21.5$

8 Evaluate $9 + 3g$ when **i** $g = -5$ **ii** $g = -4$ **iii** $g = 1.5$

9 Find the value of 2b + 4 when **i** $b = 3.5$ **ii** $b = -0.5$ **iii** $b = 0.15$

10 Evaluate $4w - 3$ when **i** $w = -2$ **ii** $w = -3$ **iii** $w = 2.5$

11 Evaluate $10 - x$ when **i** $x = -3$ **ii** $x = -6$ **iii** $x = 4.6$

12 Find the value of $5t - 1$ when **i** $t = 2.4$ **ii** $t = -2.6$ **iii** $t = 0.05$

13 Find the value of $4 + 3y$ when **i** $y = 1.5$ **ii** $y = -2.9$ **iii** $y = 0.08$

14 Evaluate $11 - 3t$ when **i** $t = 2.5$ **ii** $t = -2.8$ **iii** $t = 0.99$

HOMEWORK 3C

1 Where $P = x^2$, find P when **a** $x = 4$ **b** $x = -4$ **c** $x = 1.1$

2 Where $H = a^2 + c^2$, find H when **a** $a = 3$ and $c = 4$ **b** $a = 5$ and $c = 12$

3 Where $K = m^2 - n^2$, find K when **a** $m = 5$ and $n = 3$ **b** $m = -5$ and $n = -2$

4 Where $P = 100 - n^2$, find P when **a** $n = 7$ **b** $n = 8$ **c** $n = 9$

5 Where $D = 5x - y$, find D when **a** $x = 4$ and $y = 3$ **b** $x = 5$ and $y = -3$

6 Where $t = 50 - w$, find t when **a** $w = 64$ **b** $w = 25$ **c** $w = 100$

Give the answers to the following.

7 Where $T = y(2x + 3y)$, find T when **a** $x = 8$ and $y = 12$ **b** $x = 5$ and $y = 7$

8 Where $m = w(t^2 + w^2)$, find m when **a** $t = 5$ and $w = 3$ **b** $t = 8$ and $w = 7$

HOMEWORK 3D

Solve the following equations.

Give your answers as fractions or decimals as appropriate.

1 $5x + 4 = 13$	**2** $4x - 11 = 23$	**3** $2x - 5 = 28$	**4** $3y - 17 = 6$
5 $4a + 9 = 12$	**6** $3x + 7 = 17$	**7** $7 + 3y = 19$	**8** $9x + 5 = 13$
9 $3x - 12 = 7$	**10** $7x + 7 = 67$	**11** $2y - 8 = 5$	**12** $5x - 6 = 16$
13 $3y + 5 = 18$	**14** $9 + 4t = 12$	**15** $3 + 3f = 11$	**16** $5 + 7k = 21$
17 $5x + 8 = 15$	**18** $4m - 7 = 12$	**19** $2t - 19 = 28$	**20** $9d + 8 = 13$
21 $3x + 7 = 11$	**22** $5y - 2 = 7$	**23** $3p + 5 = 11$	**24** $6t - 5 = 4$

HOMEWORK 3E

Solve the following equations.

1 $\frac{m}{3} = 4$	**2** $\frac{k}{4} = 3$	**3** $\frac{w}{5} = 7$	**4** $\frac{x}{3} = 8$
5 $\frac{h}{7} = 5$	**6** $\frac{d}{5} = 4$	**7** $\frac{p}{2} + 5 = 7$	**8** $\frac{k}{4} - 3 = 5$
9 $\frac{g}{3} + 2 = 8$	**10** $\frac{m}{4} - 5 = 2$	**11** $\frac{f}{6} + 3 = 12$	**12** $\frac{h}{8} - 3 = 5$
13 $\frac{2h}{3} + 3 = 7$	**14** $\frac{5t}{4} - 2 = 6$	**15** $\frac{4d}{5} + 3 = 18$	**16** $\frac{3x}{4} - 1 = 8$
17 $\frac{5w}{3} + 5 = 12$	**18** $\frac{3m}{4} - 3 = 7$	**19** $\frac{8d}{7} + 3 = 2$	**20** $\frac{5g}{8} + 4 = 3$

HOMEWORK 3F

Solve the following equations.

1 $4x + 9 = 7$ **2** $5t + 6 = 2$ **3** $12 + 4x = 5$ **4** $14 + 2y = 3$

5 $7 - 5x = 11$ **6** $8 - 8t = 19$ **7** $7 - 4x = 22$ **8** $5x + 7 = 5$

9 $\frac{x+5}{3} = 2$ **10** $\frac{t+12}{2} = 5$ **11** $\frac{w-3}{5} = 3$ **12** $\frac{y-9}{2} = 3$

13 $\frac{2x-1}{3} = 5$ **14** $\frac{5t+9}{2} = 3$ **15** $\frac{4m+2}{5} = 5$ **16** $\frac{8p-7}{5} = 2$

17 $\frac{5x+19}{4} = 4$ **18** $\frac{17+2t}{9} = 1$ **19** $\frac{23+4x}{3} = 4$ **20** $\frac{8-2x}{11} = 1$

HOMEWORK 3G

Solve the following equations. Give your answers as fractions or decimals as appropriate.

1 $3(x + 6) = 15$ **2** $6(x - 5) = 30$ **3** $4(t + 3) = 20$

4 $5(4x + 3) = 45$ **5** $3(5y - 7) = 15$ **6** $4(5x + 2) = 70$

7 $3(4t - 1) = 78$ **8** $3(4t + 5) = 51$ **9** $4(6x + 5) = 10$

10 $5(4y - 1) = 47$ **11** $5(2k + 3) = 36$ **12** $5(3x + 8) = 30$

13 $3(5y - 7) = 21$ **14** $3(6t - 5) = 57$ **15** $8(2x - 7) = 60$

16 $8(3x - 4) = 12$ **17** $4(x + 7) = 7$ **18** $3(x - 5) = -24$

19 $5(t + 3) = 12$ **20** $4(3x - 13) = 7$ **21** $5(4t + 3) = 17$

22 $2(5x - 3) = -31$ **23** $4(6y - 7) = -5$ **24** $3(2x + 7) = 9$

HOMEWORK 3H

1 Without using a calculator, find the two consecutive whole numbers between which the solution to each of the following equations lies.

a $x^3 = 10$ **b** $x^3 = 50$ **c** $x^3 = 800$ **d** $x^3 = 300$

2 Find a solution to each of the following equations to 1 decimal place. Do not use the cube root on your calculator.

a $x^3 = 24$ **b** $x^3 = 100$ **c** $x^3 = 500$ **d** $x^3 = 200$

3 Find two consecutive whole numbers between which the solution to each of the following equations lies.

a $x^3 + x = 7$ **b** $x^3 + x = 55$ **c** $x^3 + x = 102$ **d** $x^3 + x = 89$

4 Find a solution to each of the following equations to 1 decimal place.

a $x^3 - x = 30$ **b** $x^3 - x = 95$ **c** $x^3 - x = 150$ **d** $x^3 - x = 333$

5 Show that $x^3 + x = 45$ has a solution between $x = 3$ and $x = 4$, and find the solution to 1 decimal place.

6 Show that $x^3 - 2x = 95$ has a solution between $x = 4$ and $x = 5$, and find the solution to 1 decimal place.

HOMEWORK 3I

1 A rectangle has an area of 200 cm^2. Its length is 8 cm longer than its width. Find, correct to 1 decimal place, the dimensions of the rectangle.

2 A gardener wants his rectangular lawn to be 15 m longer than the width, and the area of the lawn to be 800 m^2. What are the dimensions he should make his lawn? (Give your solution to 1 decimal place.)

3 A triangle has a vertical height 2 cm longer than its base length. Its area is 20 cm^2. What are the dimensions of the triangle? (Give your solution to 1 decimal place.)

4 A rectangular picture has a height 3 cm shorter than its length. Its area is 120 cm^2. What are the dimensions of the picture? (Give your solution to 1 decimal place.)

5 What are the dimensions, to 1 decimal place, of a cube that has a volume of 500 cm^3?

6 What is the length of one side of a cube with volume 44 m^3? (Give your answer to 1 decimal place.)

7 Find, correct to 1 decimal place, the solution to $x^4 = 64$

HOMEWORK 3J

Solve the following pairs of simultaneous equations

1 $5x + y = 17$, $3x + y = 11$ **2** $3x + 2y = 17$, $5x + 2y = 27$ **3** $3x - y = 7$, $5x + y = 17$ **4** $5x - 4y = 27$, $2x - 4y = 12$

★5 Two numbers x and y have a sum of 16 and a difference of 9.

a Set up a pair of simultaneous equations connecting x and y.

b Solve your equations for x and y.

HOMEWORK 3K

Solve the following pairs of simultaneous equations

1 $3x + 2y = 12$, $4x - y = 5$ **2** $4x + 3y = 37$, $2x + y = 17$ **3** $2x + 3y = 19$, $6x + 2y = 22$ **4** $5x - 2y = 14$, $3x - y = 9$

★5 Four cups of tea and three biscuits cost £3.35.
Three cups of tea and one biscuit cost £2.20
Let x be the cost of a cup of tea and y be the cost of a biscuit.

a Set up a pair of simultaneous equations connecting x and y.

b Solve your equations for x and y and find the cost of five cups of tea and four biscuits.

HOMEWORK 3L

Solve the following simultaneous equations.

1 $6x + 5y = 23$, $5x + 3y = 18$ **2** $3x - 4y = 13$, $2x + 3y = 20$ **3** $8x - 2y = 14$, $6x + 4y = 27$ **4** $5x + 2y = 33$, $4x + 5y = 23$

★**5** It costs two adults and three children £28.50 to go to the cinema.
It costs three adults and two children £31.50 to go to the Cinema.
Let the price of an adult ticket be £x and the price of a child's ticket be £y.

a Set up a pair of simultaneous equations connecting x and y.

b Solve your equations for x and y.

Homework 3M

Solve the following simultaneous equations.

1 $3x + y = 7$, $x - y = 5$

2 $5x + 3y = 11$, $3x + y = 7$

3 $3x + 2y = 9$, $2x - 3y = 19$

4 $2x - 5y = 10$, $4x - 3y = 13$

You may find a calculator useful for these questions.

5 $3x + 2y = 2$, $2x + 6y = 27$

6 $4x + 2y = 3$, $3x + 4y = 5$

7 $3x - 4y = 2.5$, $2x + 2y = 7.5$

8 $4x + y = 5$, $3x + 2y = 4$

★**9** Paul sold 50 tickets for a concert. He sold x tickets at £3 each and y tickets at £4.50 each. He collected £183.

i Write down two equations connecting x and y.

ii Solve these simultaneous equations to find how many of each kind of ticket he sold.

Homework 3N

Read each situation carefully, then make a pair of simultaneous equations in order to solve the problem.

1 A book and a CD cost £14.00 together. The CD costs £7 more than the book. How much does each cost?

2 10 second-class and six first-class stamps cost £4.24. 8 second-class and 10 first-class stamps cost £4.90. How much do I pay for 3 second-class and 4 first-class stamps?

3 6 cans of Coke and 5 chocolate bars cost £4.37. 3 cans of Coke and 2 chocolate bars cost £2.00. How much would 2 cans of Coke and a chocolate bar cost?

4 Three bags of sugar and four bags of rice weigh 12 kg. Five bags of sugar and two bags of rice weigh 13 kg. What would two bags of sugar and five bags of rice weigh?

5 Two cakes and three bags of peanuts contain 63 grams of fat. One cake and four bags of peanuts contain 64 grams of fat. How many grams of fat are there in each item?

★**6** Wath school buys Basic Scientific calculators and Graphical calculators to sell to students.
An order for 30 Basic Scientific calculators and 25 Graphical calculators came to a total of £1240. Another order for 25 Basic Scientific calculators and 10 Graphical calculators came to a total of £551.25.
Using £x to represent the cost of Basic Scientific calculators and £y to represent the cost of Graphical calculators, set up and solve a pair of simultaneous equations to find the cost of the next order which will be for 35 Basic Scientific calculators and 15 Graphical calculators.

Homework 3P

1 $y = mx + c$ — **i** Make c the subject. **ii** Express x in terms of y, m and c.

2 $v = u - 10t$ — **i** Make u the subject. **ii** Express t in terms of v and u.

3 $T = 2x + 3y$ — **i** Express x in terms of T and y. **ii** Make y the subject.

4 $p = q^2$ — Make q the subject.

5 $p = q^2 - 3$ — Make q the subject.

6 $a = b^2 + c$ — Make b the subject.

★7 A rocket is fired vertically upwards with an initial velocity of u metres per second. After t seconds the rocket's velocity, v metres per second, is given by the formula $v = u + gt$, where g is a constant.

a Calculate v when $u = 120$, $g = -9.8$ and $t = 6$.

b Rearrange the formula to express t in terms of v, u, and g.

c Calculate t when $u = 100$, $g = -9.8$ and $v = 17.8$.

Chapter 4 Pythagoras and trigonometry

Homework 4A

In each of the following triangles, find the hypotenuse, rounding off to a suitable degree of accuracy.

1

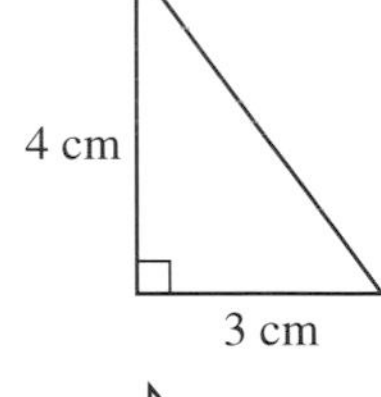

2

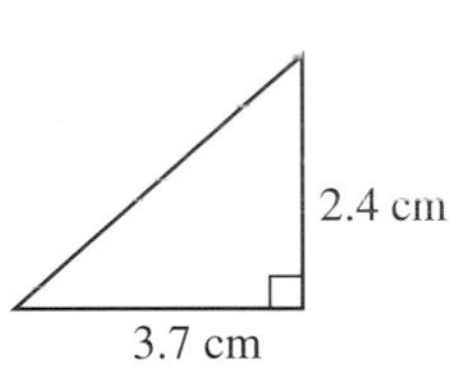

3

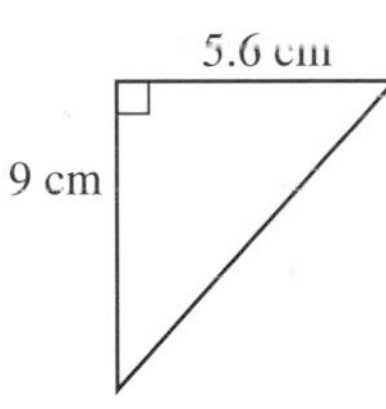

4

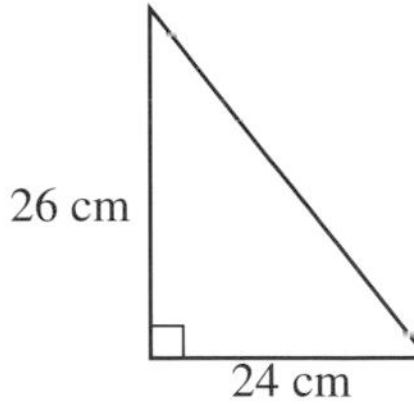

5

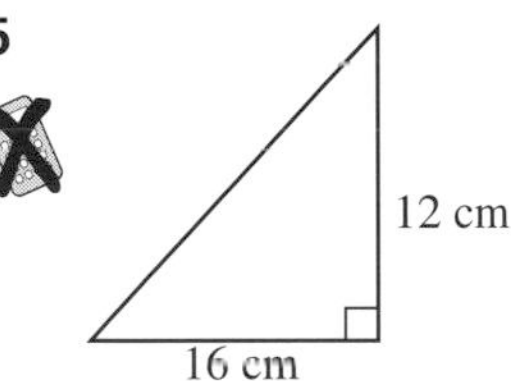

6

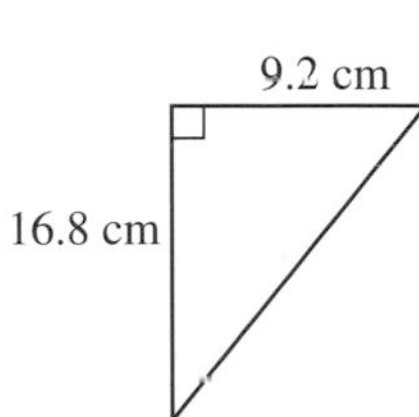

★7 This diagram is not drawn to scale. It shows the cross-section of a swimming pool 50 m long. It is 3.5 m deep at the deep end. The deepest part of the pool is 10 m long.

a Calculate the length of the sloping bottom of the pool AB.

b The pool is 20 m wide. What is its volume?

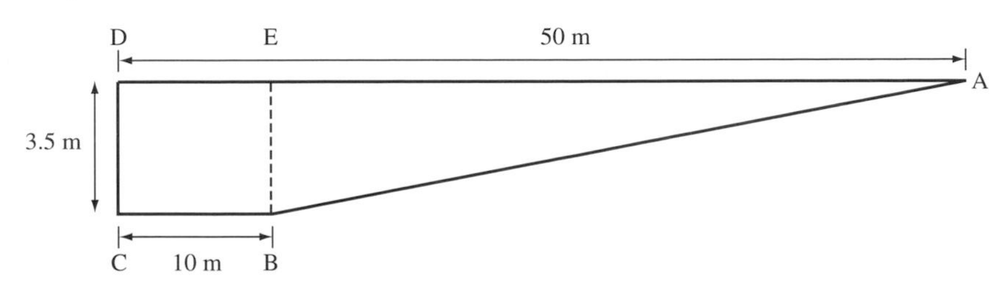

HOMEWORK 4B

1 In each of the following triangles, find the length of x to a suitable degree of accuracy.

a

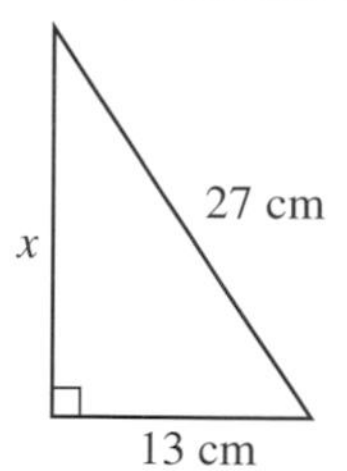

b

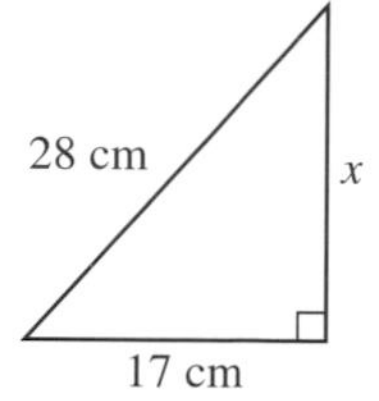

c

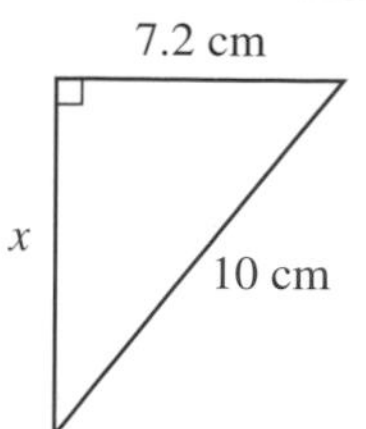

d

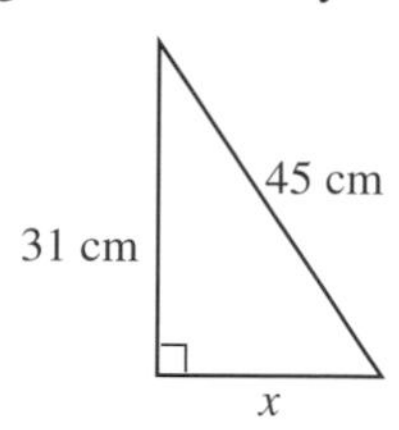

e

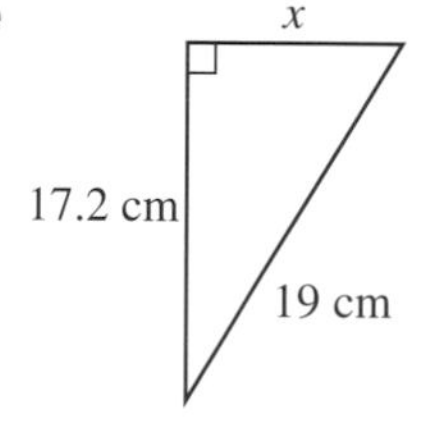

f

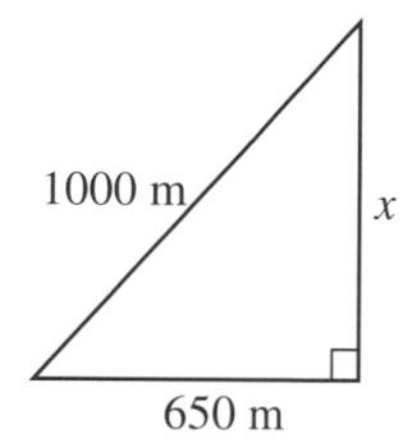

g

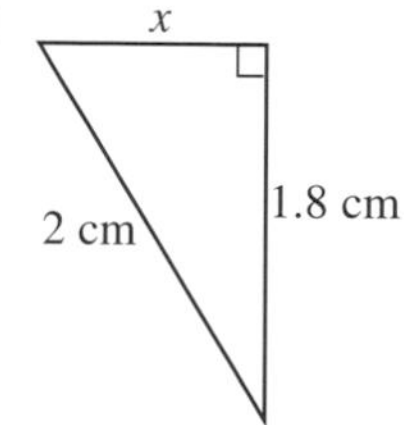

h

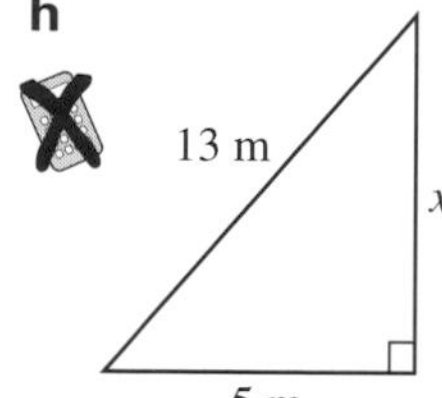

2 In each of the following triangles, find the length of x to a suitable degree of accuracy.

a

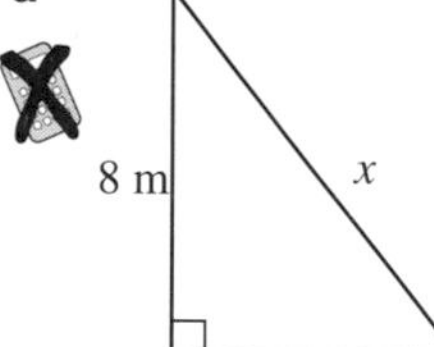

b

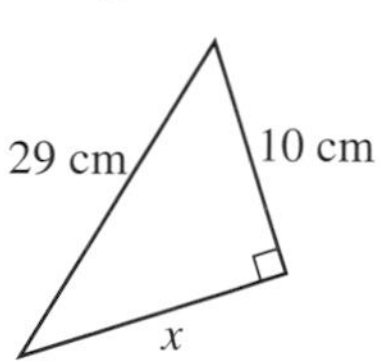

c

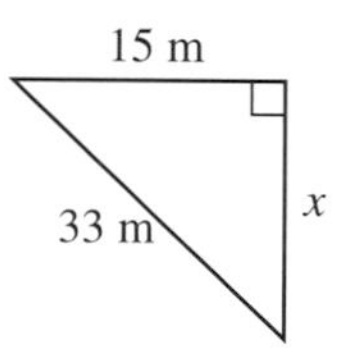

d

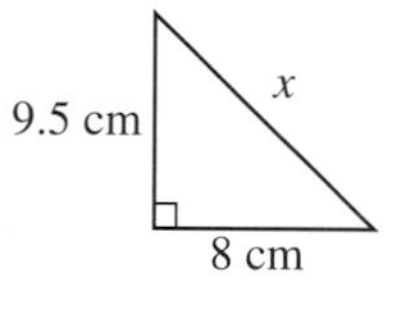

★3 The diagram shows the end view of the framework for a sports arena stand. Calculate the length AB.

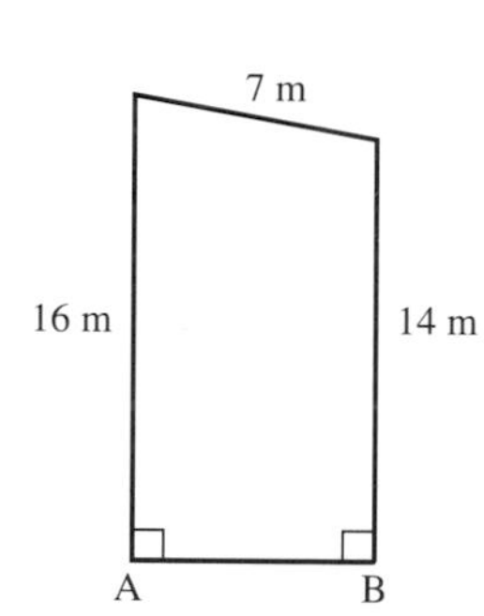

HOMEWORK 4C

1 A ladder, 15 metres long, leans against a wall. The ladder reaches 12 metres up the wall. How far away from the foot of the wall is the foot of the ladder.

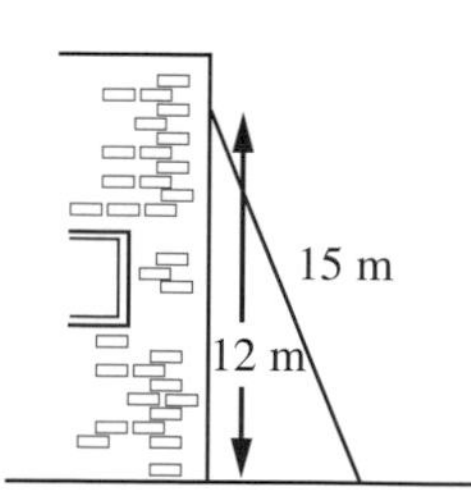

2 A rectangle is 3 metres long and 1.2 m wide.
How long is the diagonal?

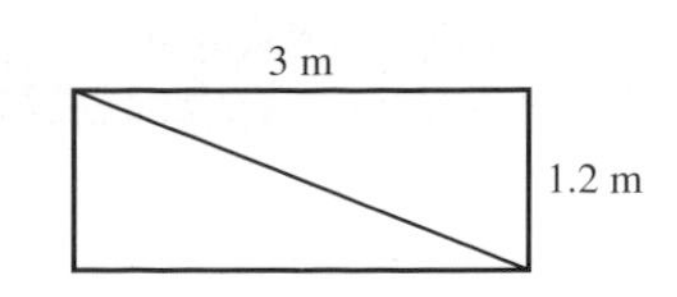

3 How long is the diagonal of a square with a side of 10 metres?

4 A ship going from a port to a lighthouse steams 8 km east and 6 km north. How far is the lighthouse from the port?

5 At the moment, three towns, A, B and C, are joined by two roads, as in the diagram. The council wants to make a road which runs directly from A to C. How much distance will the new road save?

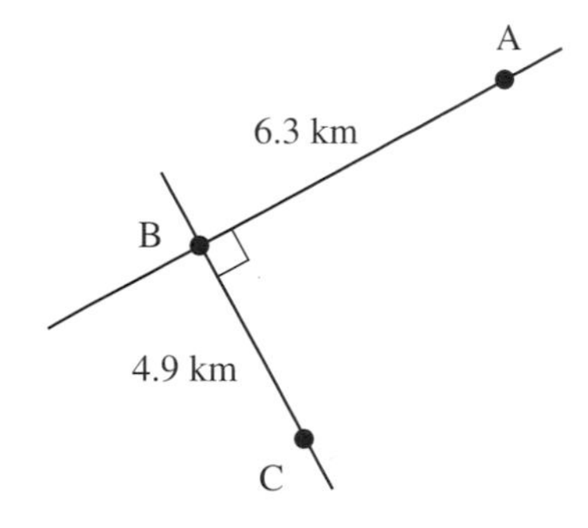

6 An 8-metre ladder is put up against a wall.

a How far up the wall will it reach when the foot of the ladder is 1 m away from the wall?

b When it reaches 7 m up the wall, how far is the foot of the ladder away from the wall?

7 How long is the line that joins the two co-ordinates A(1, 3) and B(2, 2)?

8 A rectangle is 4 cm long. The length of its diagonal is 5 cm. What is the area of the rectangle?

9 Is the triangle with sides 11 cm, 60 cm and 61 cm a right-angled triangle?

10 How long is the line that joins the two co-ordinates A(–3, –7), and B(4, 6)?

★11 The diagram shows a voyage from A to position B. The boat sails due east from A for 27 km to position C. The boat then changes course and sails for 30 km to position B. On a map, the distance between A and C is 10.8 cm.

a Work out the scale of the map.

b Calculate the distance, in km, of B from A.

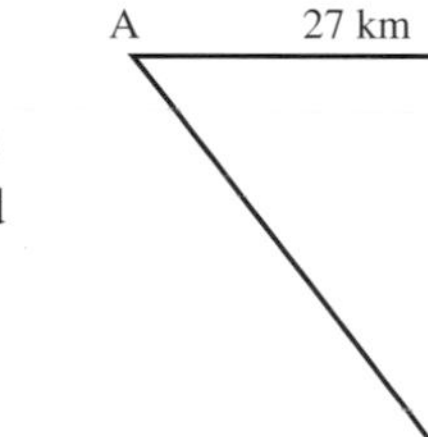

HOMEWORK 4D

1 Calculate the area of these isosceles triangles.

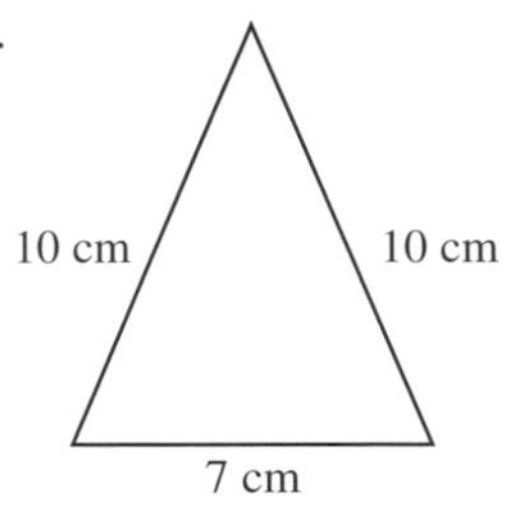

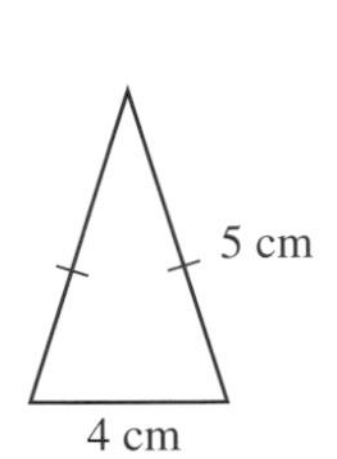

2 Calculate the area of an isosceles triangle whose sides are 10 cm, 10 cm and 8 cm.

3 Calculate the area of an equilateral triangle of side 10 cm.

4 **a** Calculate the area of an equilateral triangle of side 20 cm.
b Explain why the answer to **4a** is not twice that of Question **3**.

5 An isosceles triangle has sides of 6 cm and 8 cm.
a Sketch the two isosceles triangles that fit this data.
b Which of the two triangles has the greater area?

★6 The diagram shows an isosceles triangle of base 10 mm and side 12 mm. Calculate the area of the triangle.

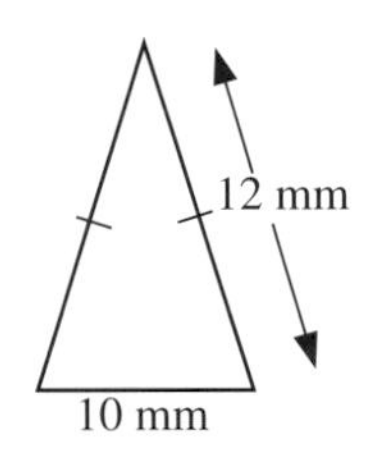

HOMEWORK 4E

1 A box measures 6 cm by 8 cm by 10 cm.
a Calculate the lengths of
i AC **ii** BG **iii** BE
b Calculate the diagonal distance BH.

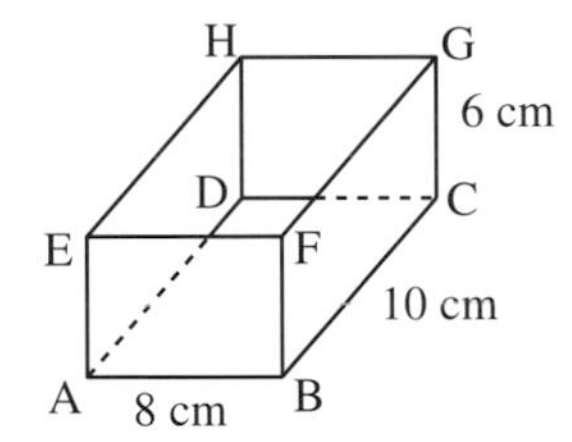

2 A garage is 5 m long, 5 m wide and 2 m high. Can a 7 m long pole be stored in it?

3 Spike, a spider, is at the corner S of the wedge shown in the diagram. Fred, a fly, is at the corner F of the same wedge.
a Calculate the two distances Spike would have to travel to get to Fred if she used the edges of the wedge.
b Calculate the distance Spike would have to travel across the face of the wedge to get directly to Fred.

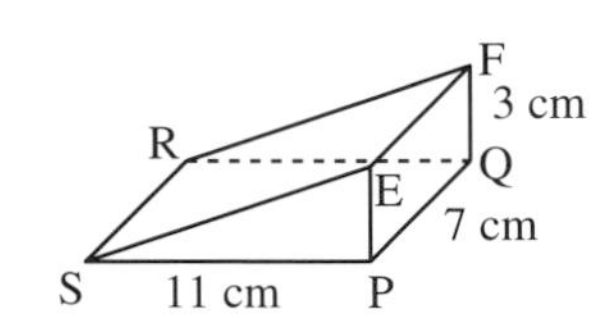

4 A corridor is 5 m wide and turns through a right angle, as in the diagram. What is the longest pole that can be carried along the corridor horizontally? If the corridor is 3 m high, what is the longest pole that can be carried along in any direction?

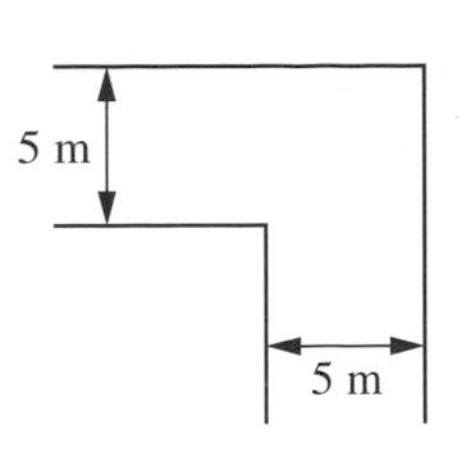

5 For the box shown on the right, find the lengths of

a DG
b HA
c DB
d AG

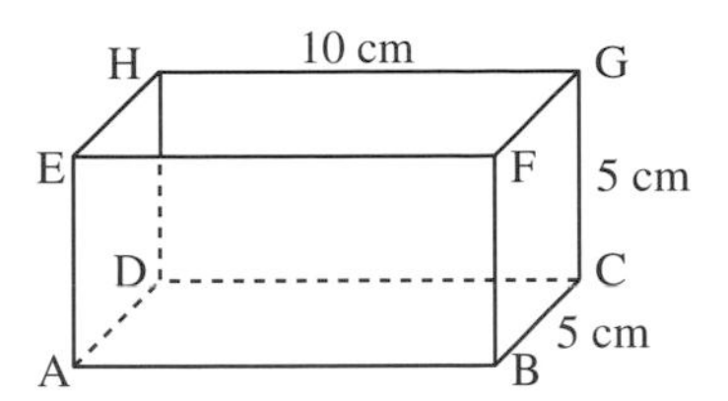

6 The diagram shows a square-based pyramid with base length 7 cm and sloping edges 12 cm. M is the mid-point of the side AB, X is the mid-point of the base, and E is directly above X.

a Calculate the length of the diagonal AC.
b Calculate EX, the height of the pyramid.
c Using triangle ABE, calculate the length EM.

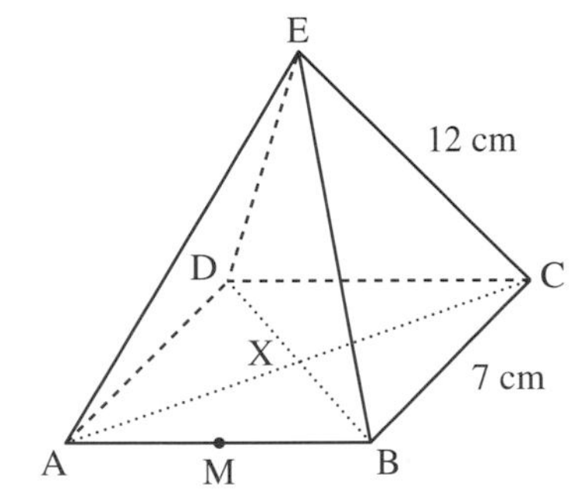

★7 **a** Is the triangle with sides 9, 40, 41 cm a right-angled triangle?

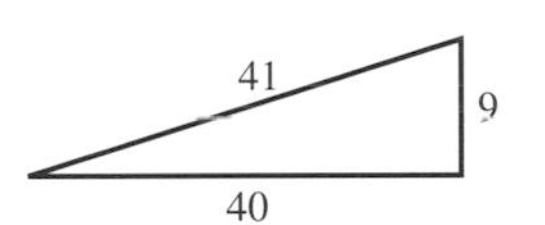

b Find the length of the diagonal AB of the cuboid 9 cm by 9 cm by 40 cm.

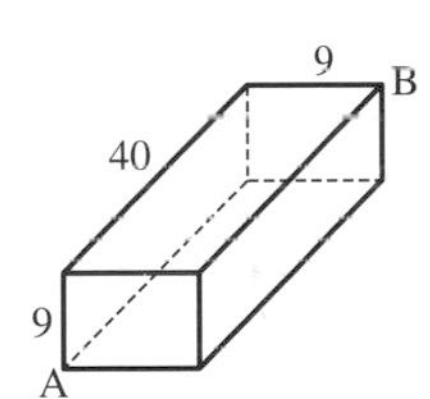

HOMEWORK 4F

1 Find these values, rounding off your answers to 3 sf.

a sin 52° **b** sin 46° **c** sin 76.3° **d** sin 90°

2 Find these values, rounding off your answers to 3 sf.

a cos 52° **b** cos 46° **c** cos 76.3° **d** cos 90°

3 **a** Calculate $(\sin 52°)^2 + (\cos 52°)^2$ **b** Calculate $(\sin 46°)^2 + (\cos 46°)^2$

c Calculate $(\sin 76.3°)^2 + (\cos 76.3°)^2$ **d** Calculate $(\sin 90°)^2 + (\cos 90°)^2$

e What do you notice about your answers?

4 Use your calculator to work out the value of

a tan 52° **b** tan 46° **c** tan 76.3° **d** tan 0°

5 Use your calculator to work out the value of

a sin 52° ÷ cos 52° **b** sin 46° ÷ cos 46° **c** sin 76.3° ÷ cos 76.3°

d sin 0° ÷ cos 0°

e What connects your answers with the answers to Question **4**?

6 Use your calculator to work out the value of

a $6 \sin 55°$ **b** $7 \cos 45°$ **c** $13 \sin 67°$ **d** $20 \tan 38°$

7 Use your calculator to work out the value of

a $\dfrac{6}{\sin 55°}$ **b** $\dfrac{7}{\cos 45°}$ **c** $\dfrac{13}{\sin 67°}$ **d** $\dfrac{20}{\tan 38°}$

8 Using the following triangle, calculate sin, cos, and tan for the angle marked x. Leave your answers as fractions.

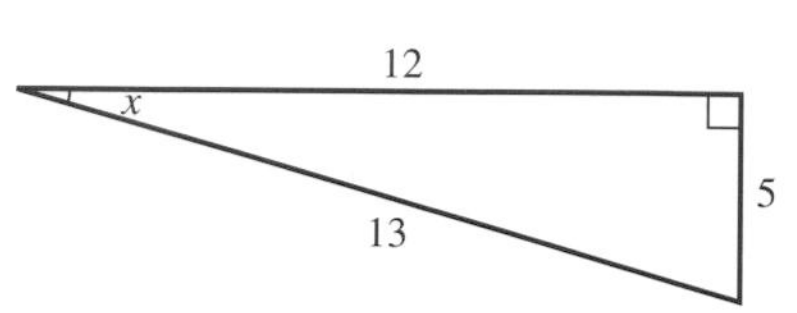

★9 You are given that $\sin x = \dfrac{5}{\sqrt{34}}$. Work out the value of $\tan x$.

HOMEWORK 4G

Use your calculator to find the answers to the following. Give your answers to 1 dp.

1 What angles have sines of

a 0.4 **b** 0.707 **c** 0.879 **d** 0.666666666666666…

2 What angles have cosines of

a 0.4 **b** 0.707 **c** 0.879 **d** 0.333333333333333…

3 What angles have tangents of

a 0.4 **b** 1.24 **c** 0.875 **d** 2.625

4 What angles have sines of

a $3 \div 8$ **b** $1 \div 3$ **c** $3 \div 10$ **d** $5 \div 8$

5 What angles have cosines of

a $3 \div 8$ **b** $1 \div 3$ **c** $3 \div 10$ **d** $5 \div 8$

6 What angles have tangents of

a $3 \div 8$ **b** $3 \div 2$ **c** $5 \div 7$ **d** $19 \div 5$

7 You are given that $\sin 54° = 0.809$ to 3 dp. What angle has a cosine of 0.809?

HOMEWORK 4H

1 Find the value marked x in each of these diagrams.

a

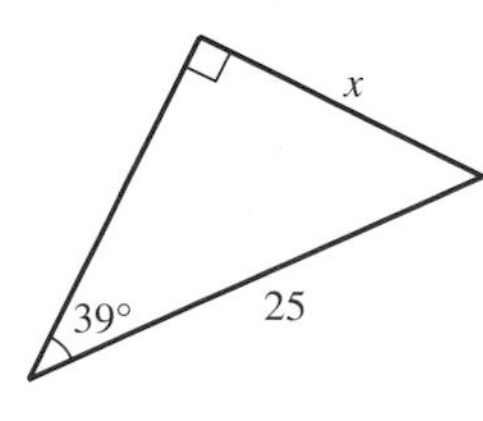

b

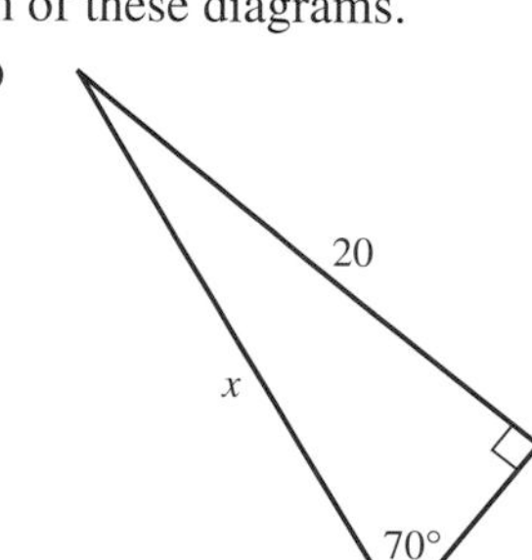

c

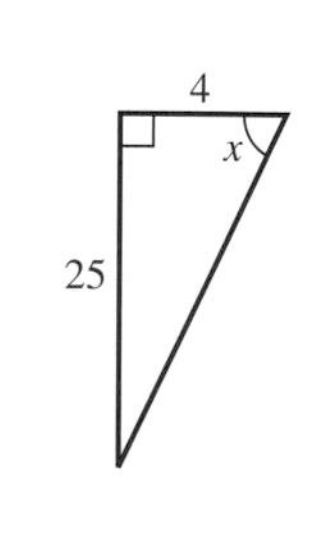

d 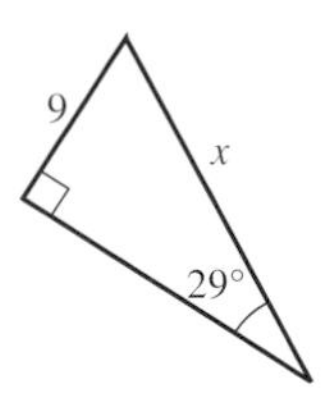

e 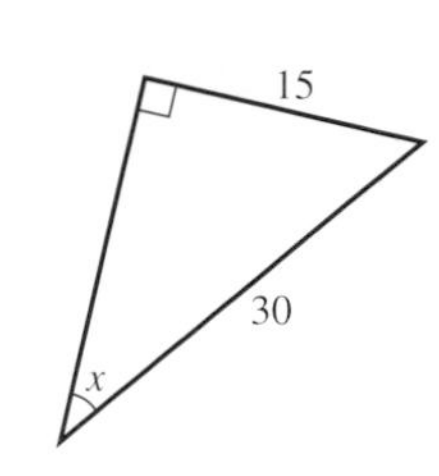

f 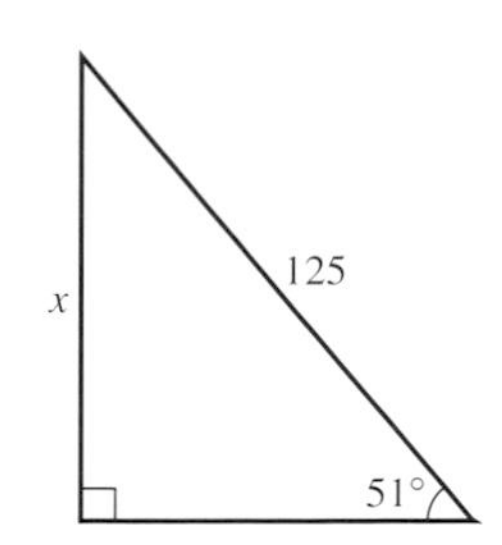

2 Angle θ has a sine of $\frac{7}{20}$. Calculate the missing lengths in these triangles.

a 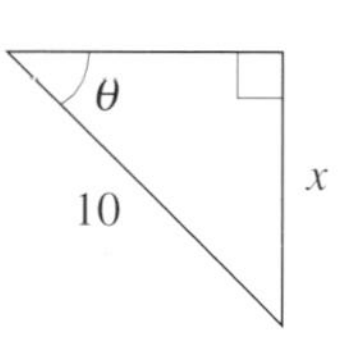

b 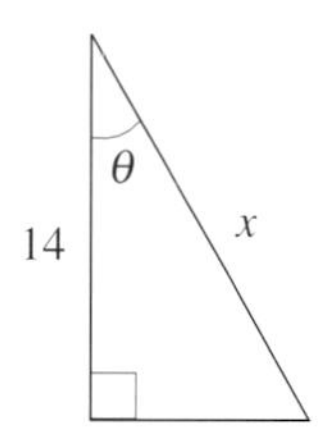

c 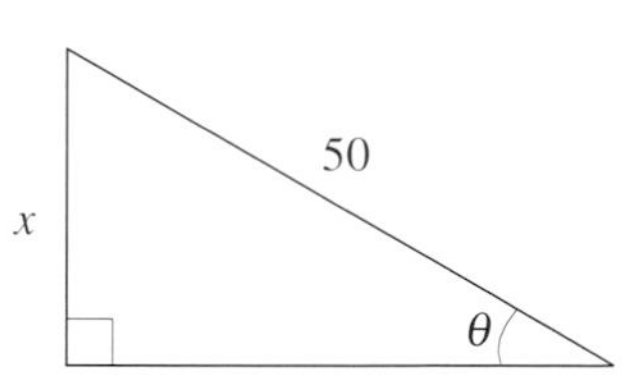

★3 Caxton is due north of Ashville and due west of Peaton. A pilot flies directly from Ashville to Peaton, a distance of 15 km, on a bearing of 050°.

a Calculate the direct distance from Caxton to Peaton.

b Find the bearing of Ashville from Peaton.

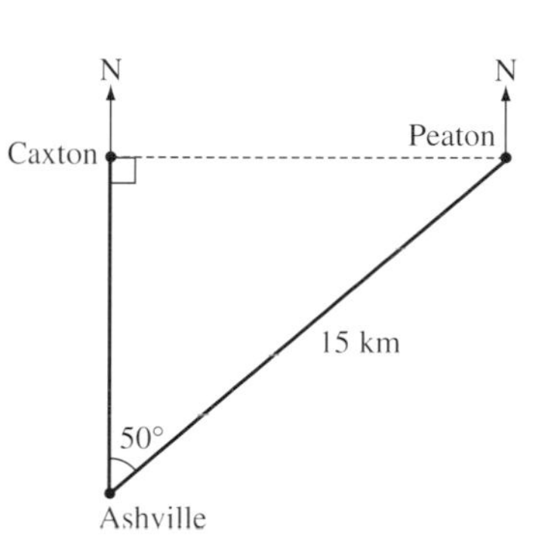

Homework 4I

1 Find the value marked x in each of these triangles.

a 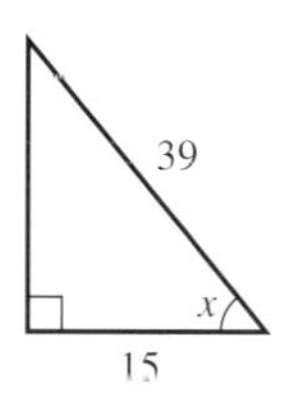

b 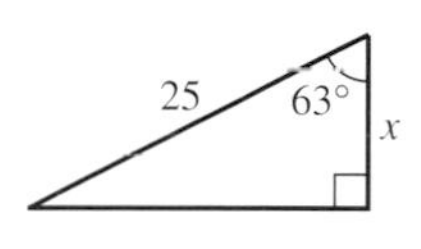

c 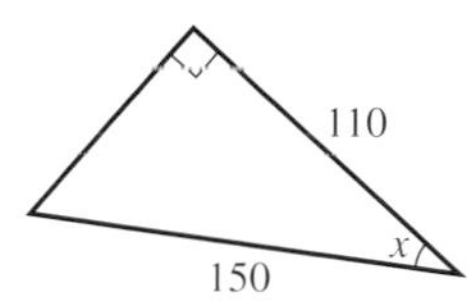

d 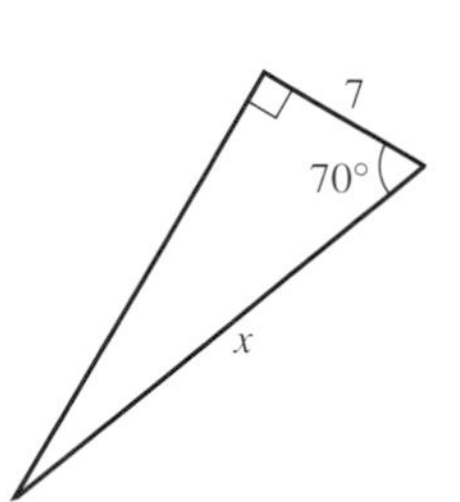

e 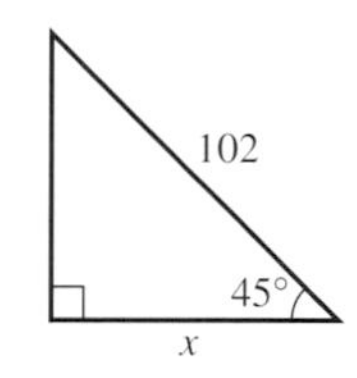

f 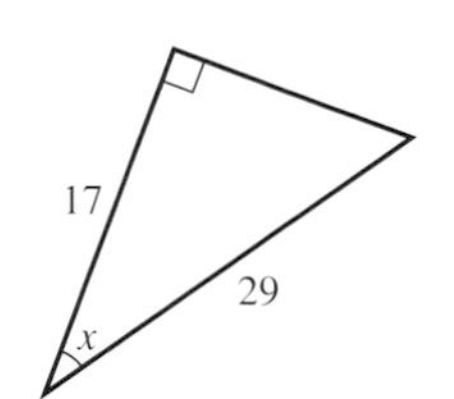

2 Angle θ has a cosine of $\frac{7}{15}$. Calculate the missing lengths in these triangles.

a

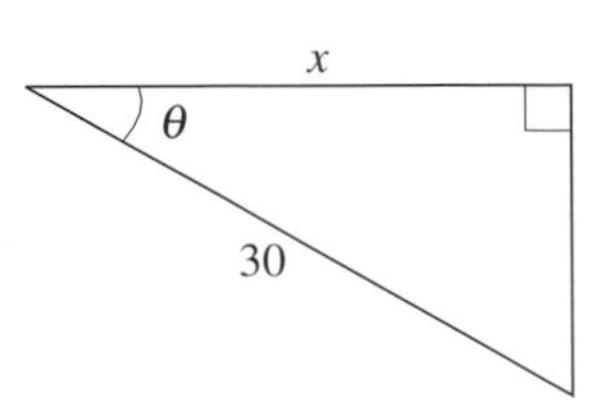

b

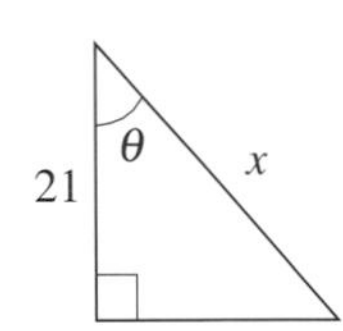

c

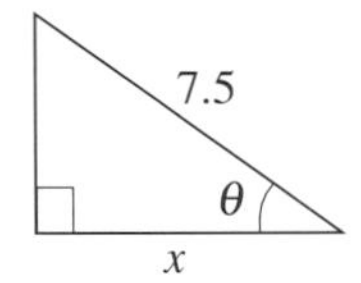

★3 The diagram shows the positions of three telephone masts A, B and C.
Mast C is 6 kilometres due East of Mast B.
Mast A is due North of Mast B, and 9 kilometres from Mast C.

a Calculate the distance of A from B. Give your answer in kilometres, correct to 3 sf.

b Calculate the size of the angle marked $x°$. Give your angle correct to 1 dp.

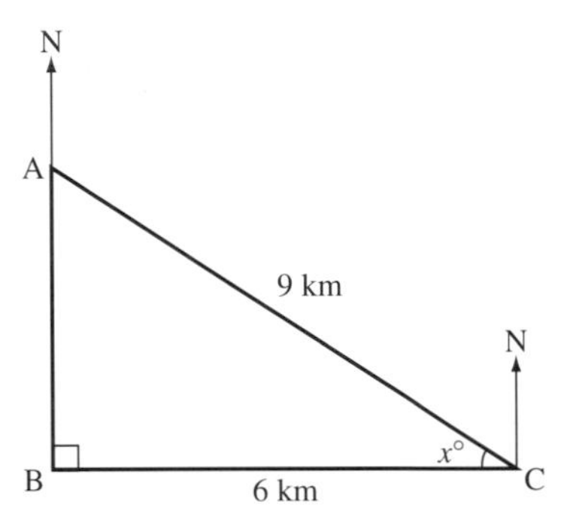

HOMEWORK 4J

1 Find the value marked x in each of these triangles.

a

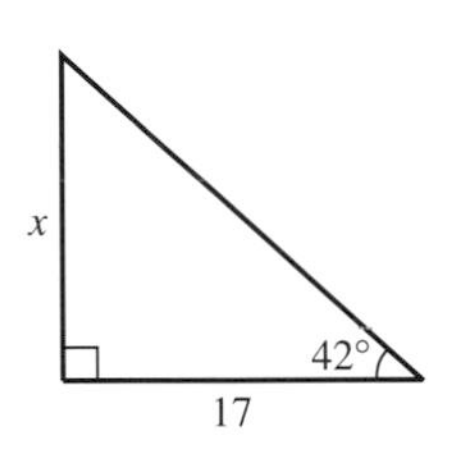

b

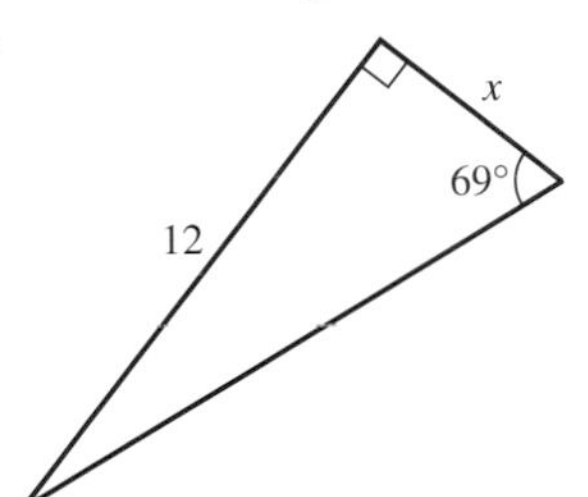

c

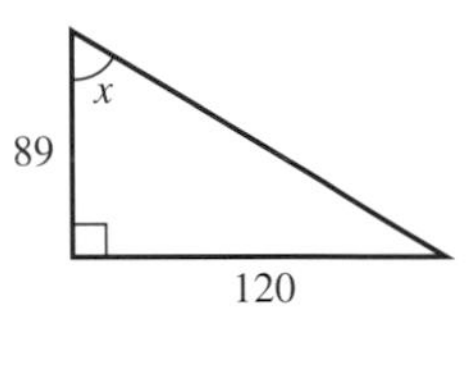

d

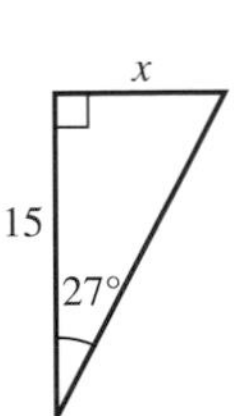

e

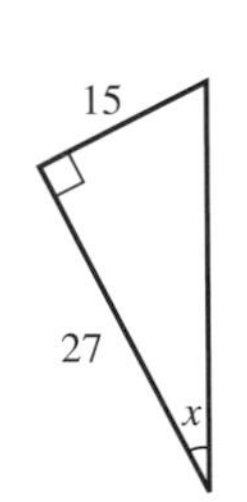

f

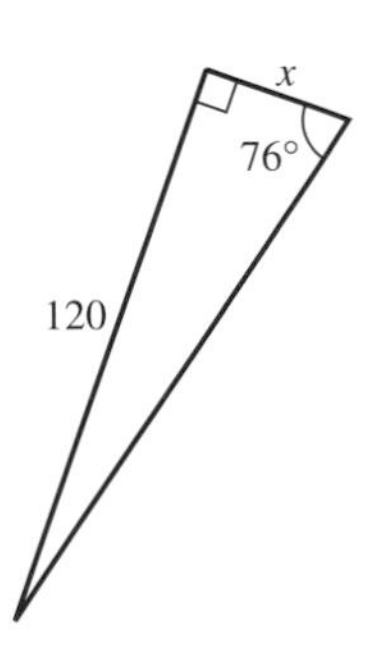

2 Angle θ has a tangent of $\frac{2}{3}$. Calculate the missing lengths in these triangles.

a

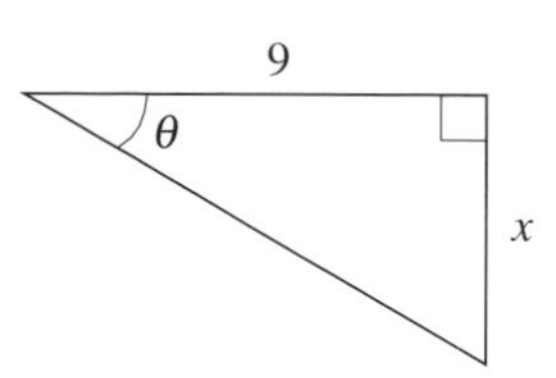

b

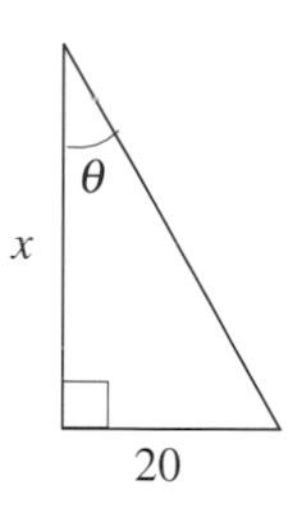

c

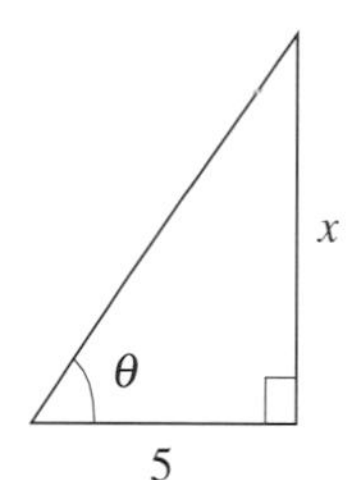

★3 The sensor for a security light is fixed to a house wall 2.25 m above the ground. It can detect movement on the ground up to 15 m away from the house. B is the furthest point where the sensor, A, can detect movement. Calculate the size of angle x.

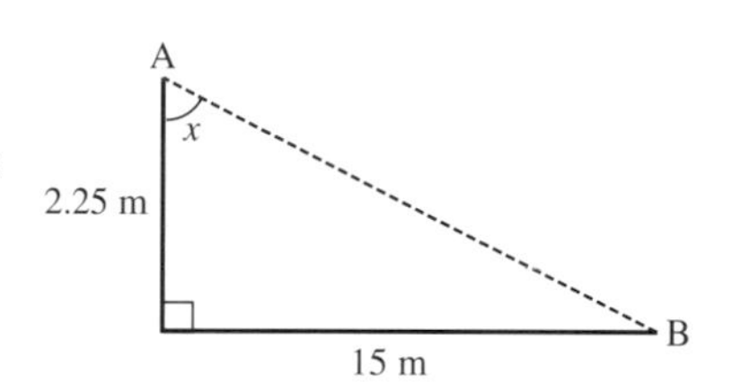

HOMEWORK 4K

1 Find the angle or length marked x in each of these triangles.

a

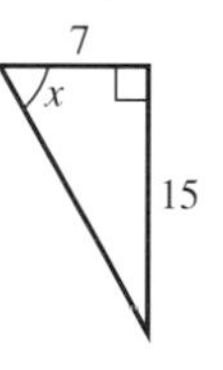

b

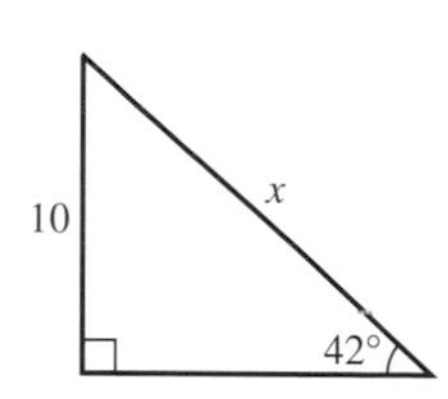

c

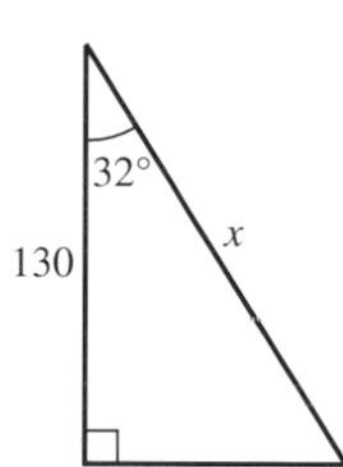

d

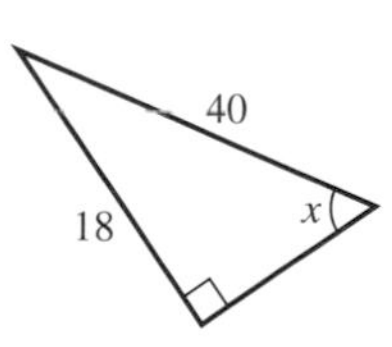

e

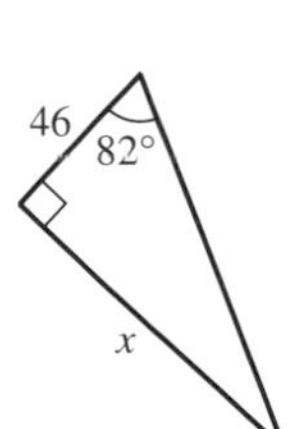

f

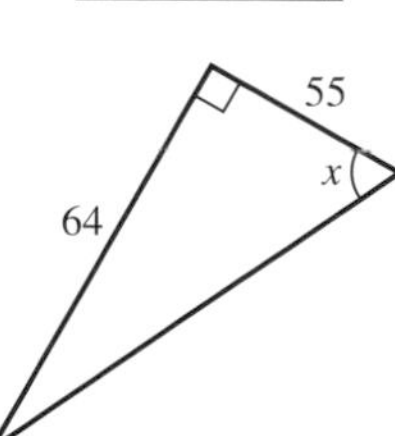

g

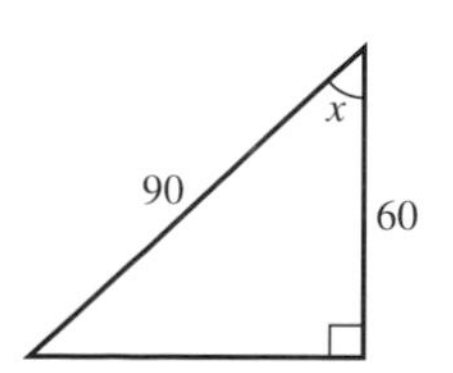

h

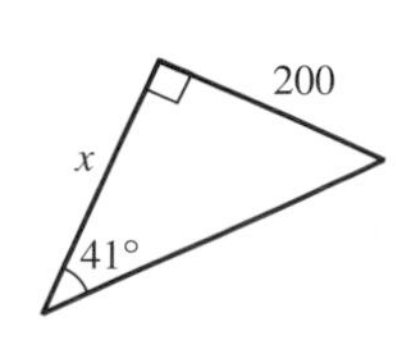

i

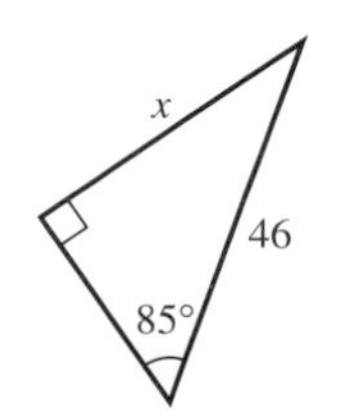

 ★2 The diagram shows a right-angled triangle, ABC.
Angle C = 90° and AB = 10 cm.
Given that cos B = 0.8, sin B = 0.6 and tan B = 0.75, calculate the length of AC.

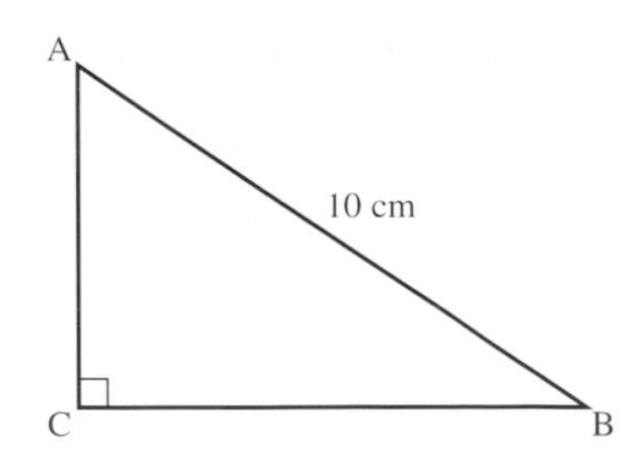

3 A lift at the seaside takes people from sea level to the top of a cliff, as shown. From sea level to the top of the cliff, the lift travels 23 m and rises a height of 21 m.

a Calculate the distance AC. Give your answer to an appropriate degree of accuracy.

b Calculate angle BCA.

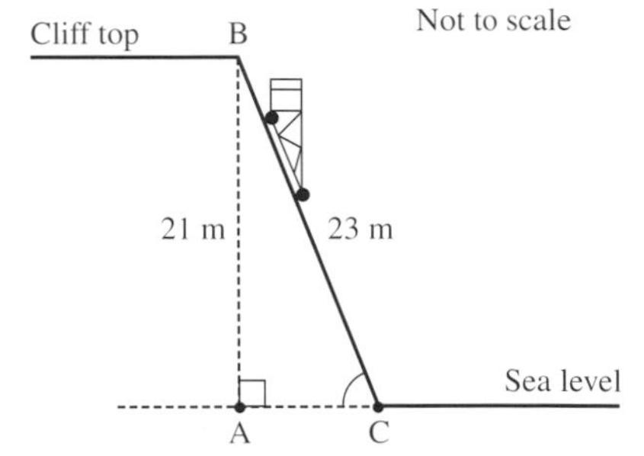

HOMEWORK 4L

1 A ladder 8 m long rests against a wall. The foot of the ladder is 2.7 m from the base of the wall. What angle does the ladder make with the ground?

2 The ladder in Question **1** has a 'safe angle' with the ground of between 70° and 80°. What are the safe limits for the distance of the foot of the ladder from the wall?

3 Angela walks 60 m from the base of a block of flats and then measures the angle from the ground to the top of the flats to be 42° as shown in the diagram. How high is the block of flats?

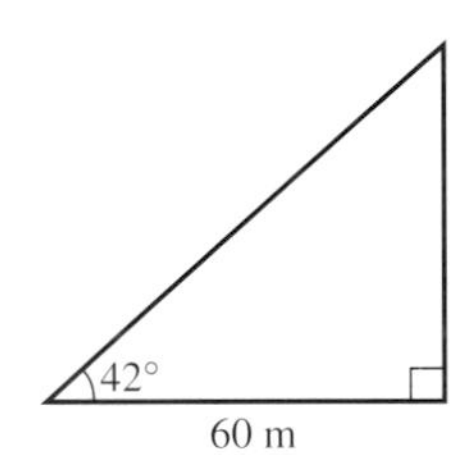

4 A slide is at an angle of 46° to the horizontal.
The slide is 7 metres long. How high is the top of the slide above the ground?

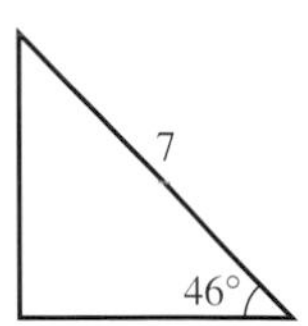

5 Use trigonometry to calculate the angle that the diagonal makes with the long side of a rectangle 9 cm by 5 cm.

★6 The diagram shows the end view of a . building BCD is a right-angled triangle. Angle BCD = 90°. BC = 5 m and BD = 9.5 m.

a Calculate angle CBD.

CH is perpendicular to AE.
AB = ED = 3.5.

b Calculate CH, the height of the building.

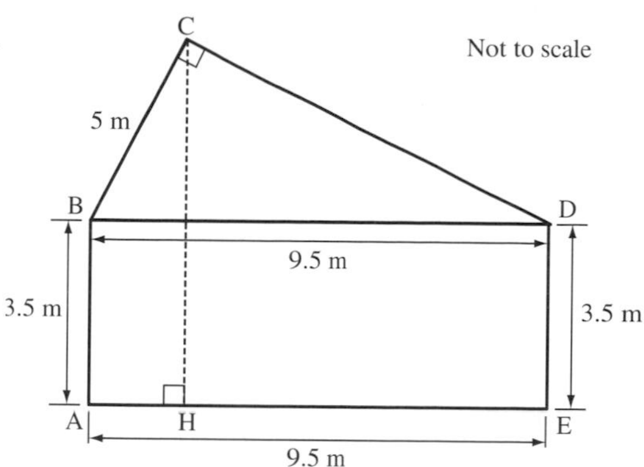

Homework 4M

1 Eric sees an aircraft in the sky. The aircraft is at a horizontal distance of 15 km from Eric. The angle of elevation is 42°. How high is the aircraft?

2 A man standing 100 m from the base of a block of flats looks at the top of it and notices that the angle of elevation of the top is 49°. How high are the flats?

3 A man stands 15 metres from a tree. The angle of elevation of the top of the tree from his eye is 25°. If his eye is 1.5 metres above the ground, how high is the tree?

4 A bird, sat on the very top of the tree in Question **3**, sees a worm just by the foot of the man. What is the angle of depression from the bird's eye to the worm?

5 I walk 200 metres away from a chimney that is 120 metres high. What is the angle of elevation from my eye to the top of the chimney? (Ignore the height of eye above ground).

6 If you are now told that the height of the eye in Question **5** was 1.8 metres above ground, how much different is the angle of elevation?

★7 A boat B is moored 50 m from the foot of a vertical cliff. The angle of depression of the boat from the top of the cliff is 52°.

a Calculate the height of the cliff.

b The boat is released from its mooring and it drifts 350 m directly away from the cliff. Calculate the angle of elevation of the top of the cliff from the boat.

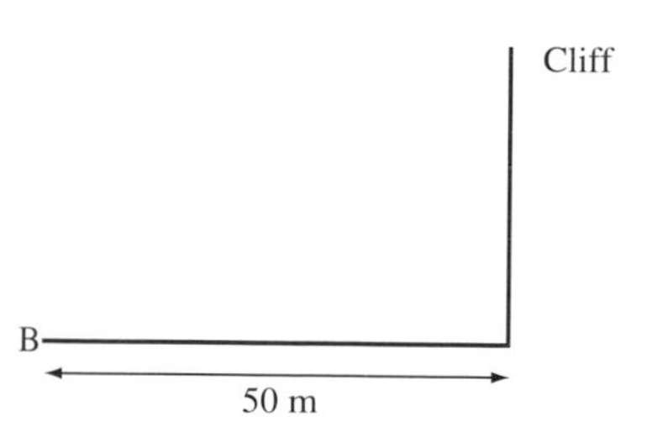

Homework 4N

1 **a** A ship sails for 85 km on a bearing of 067°. How far east has it travelled?
b How far north has the ship sailed?

2 Rotherham is 11 miles south of Barnsley and 2 miles west of Barnsley. What is the bearing of

a Barnsley from Rotherham **b** Rotherham from Barnsley?

3 A plane sets off from airport A and flies due east for 100 km, then turns to fly due south for 80 km before landing at an airport B. What is the bearing of airport B from airport A?

4 Mountain A is due east of a walker. Mountain B is due south of the walker. The guidebook says that mountain A is 5 km from mountain B, on a bearing of 038°. How far is the walker from mountain B?

5 The diagram shows the relative distances and bearings of three ships A, B and C.

a How far north of A is B? (Distance x on diagram.)

b How far north of B is C? (Distance y on diagram.)

c How far west of A is C? (Distance z on diagram.)

d What is the bearing of A from C? (Angle $w°$ on diagram.)

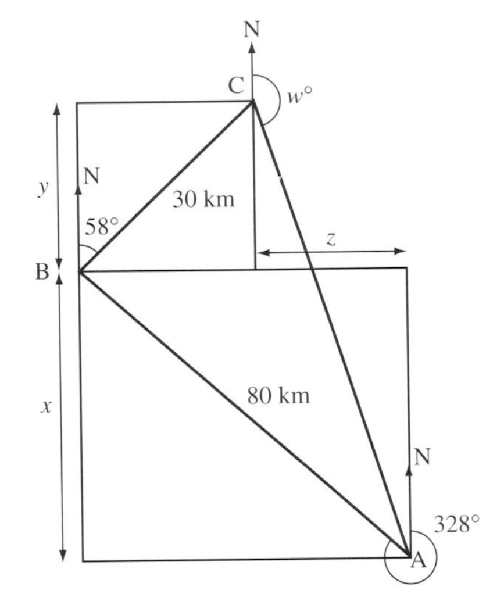

★6 An aeroplane is flying from Leeds (L) to London Heathrow (H). It flies 150 miles on a bearing 136° to point A. It then turns through 90° and flies the final 80 miles to H.

a **i** Show clearly why the angle marked x is equal to 46°.

ii Give the bearing of H from A.

b Use Pythagoras' theorem to calculate the distance LH.

c **i** Calculate the size of the angle marked y.

ii Work out the bearing of L from H.

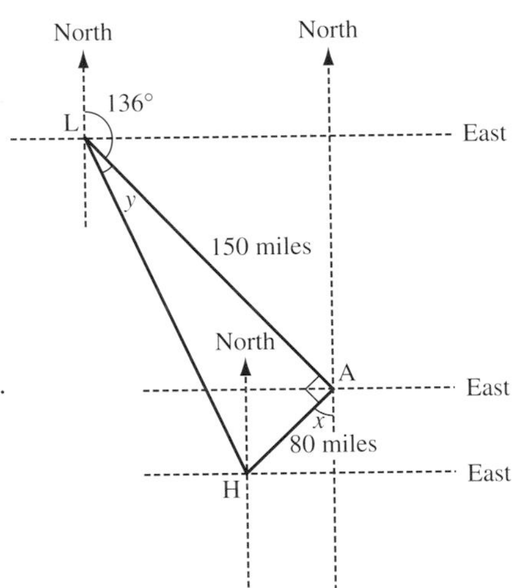

HOMEWORK 4P

In Questions **1** and **2**, find the side or angle marked x.

1

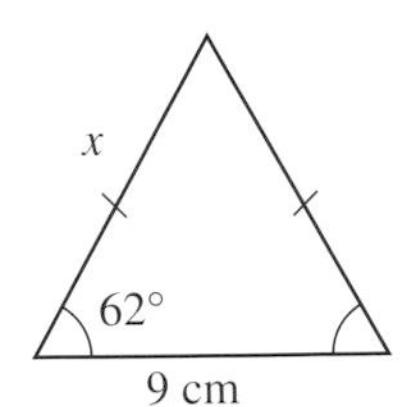

2

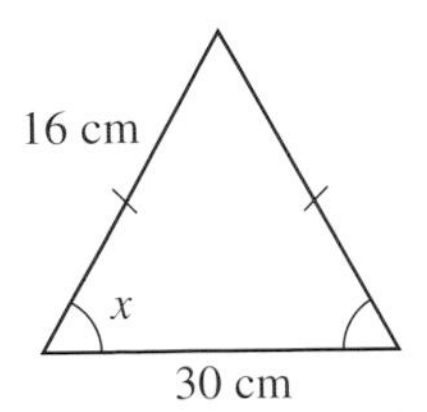

3 This diagram below shows a roof truss. How wide is the roof?

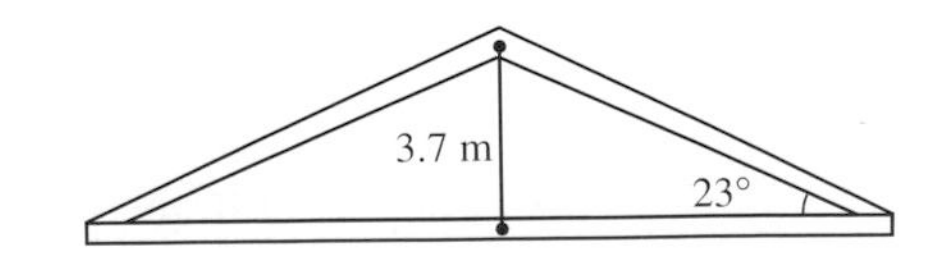

4 Calculate the area of each of these triangles.

a

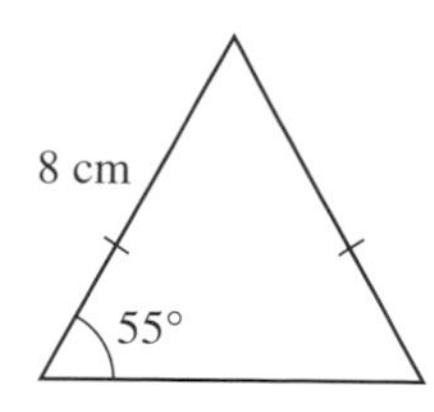

b

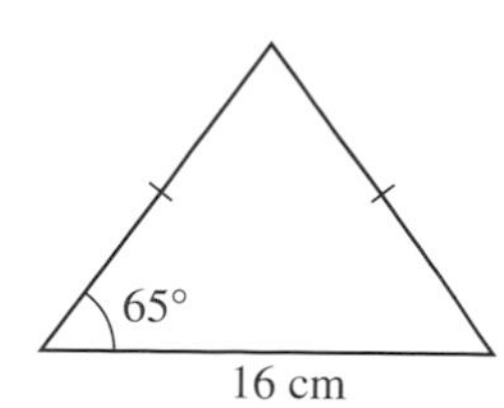

 **★5** An isosceles triangle has two sides of 12 cm and an angle of 62°. Calculate both possible areas.

Chapter 5 Geometry

Homework 5A

1 Calculate the sum of the interior angles of polygons with

a 7 sides **b** 11 sides **c** 20 sides **d** 35 sides

2 Calculate the size of the interior angle of regular polygons with

a 15 sides **b** 18 sides **c** 30 sides **d** 100 sides

3 Find the number of sides of the polygon with the interior angle sum of

a 1440° **b** 2520° **c** 6120° **d** 6840°

 **4** Find the number of sides of the regular polygon with an exterior angle of

a 20° **b** 30° **c** 18° **d** 4°

 **5** Find the number of sides of the regular polygon with an interior angle of

a 135° **b** 165° **c** 170° **d** 156°

 **6** What is the name of the regular polygon whose interior angle is treble its exterior angle?

★7 Anne measured all the interior angles in a polygon. She added them up to make 1325°, but she had missed out one angle. What is the

a name of the polygon that Anne measured

b size of the missing angle?

Homework 5B

 **1** For each of these shapes, calculate the value of the lettered angles.

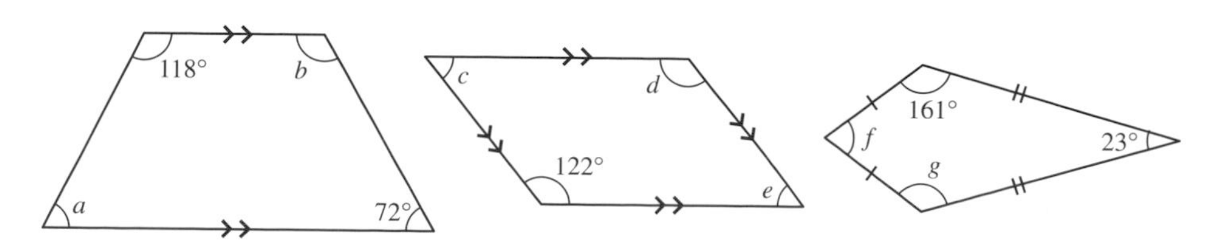

2 Calculate the values of x and y in each of these shapes.

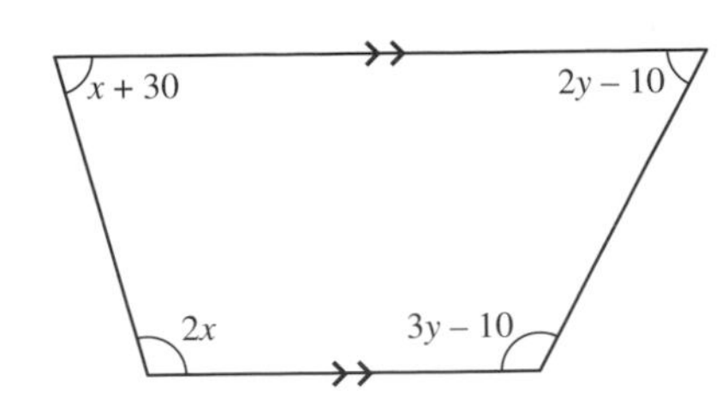

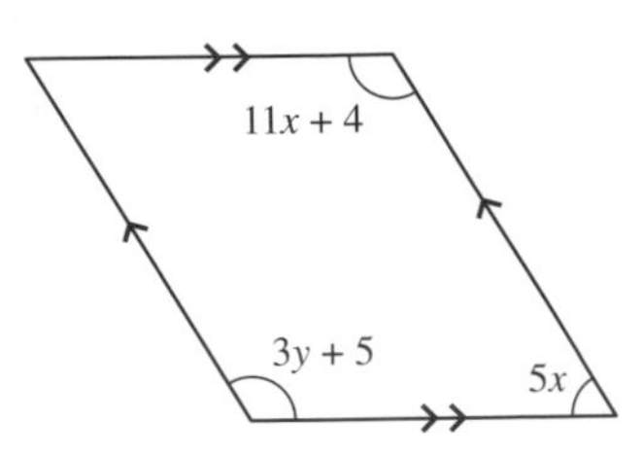

★3 Find the value of x in each of these quadrilaterals with the following angles and hence state the type of quadrilateral it is.

a $x + 10°, x + 30°, x - 30°, x - 50°$ **b** $x°, x - 10°, 3x - 15°, 3x - 15°$

PROOF **4** **a** What do the Interior angles of a quadrilateral add up to?

b Use the fact that the angles of a triangle add up to 180° to prove that the sum of the interior angles of any quadrilateral is 360°.

HOMEWORK 5C

1 Find the value of x in each of these circles with centre O.

a

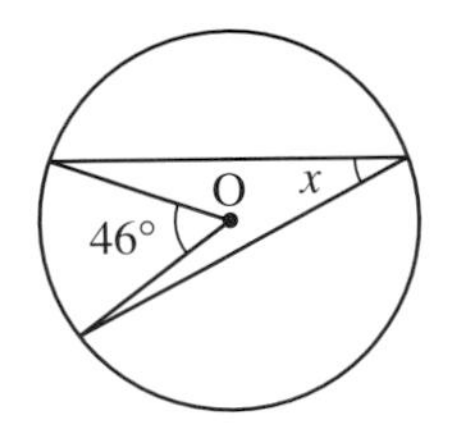

b

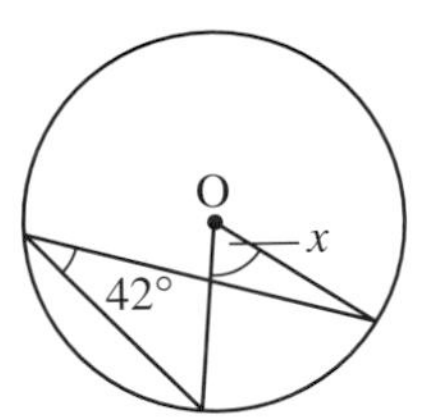

c

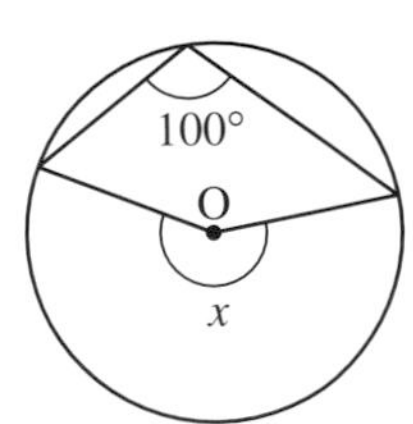

d

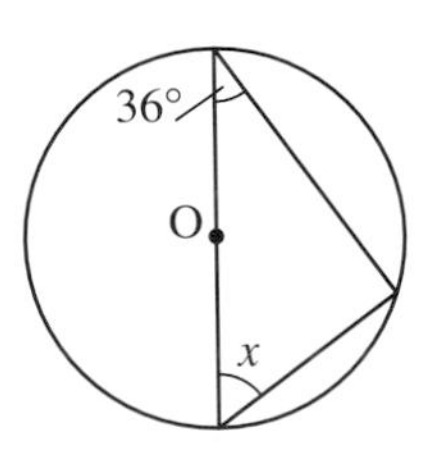

e

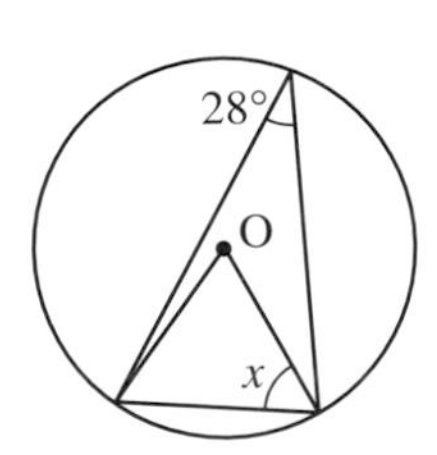

f

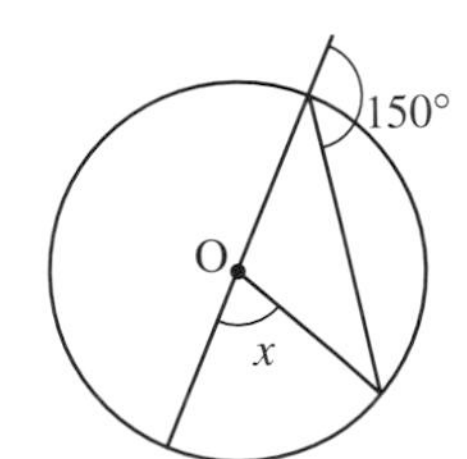

2 Find the value of x in each of these circles.

a

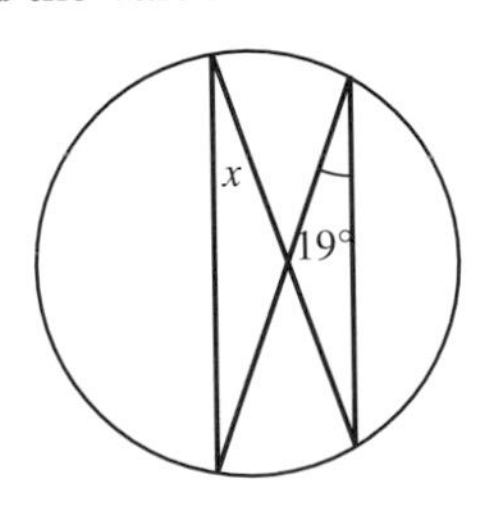

b

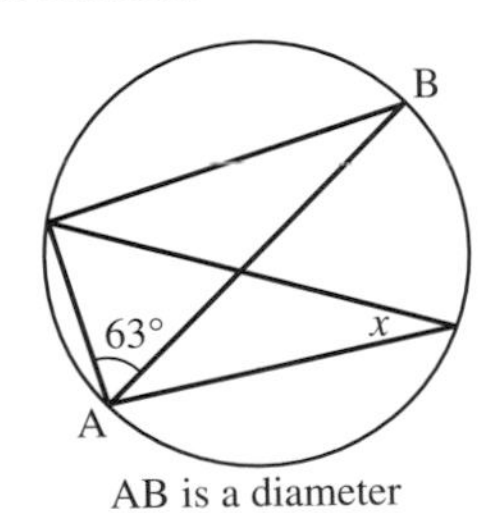

AB is a diameter

c

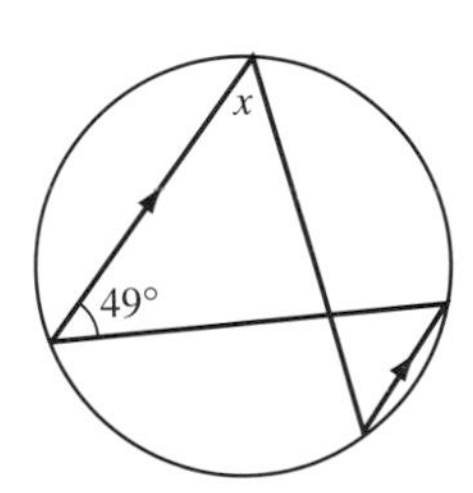

3 In the diagram, O is the centre of the circle. Find

a ∠EDF

b ∠DEG

c ∠EGF

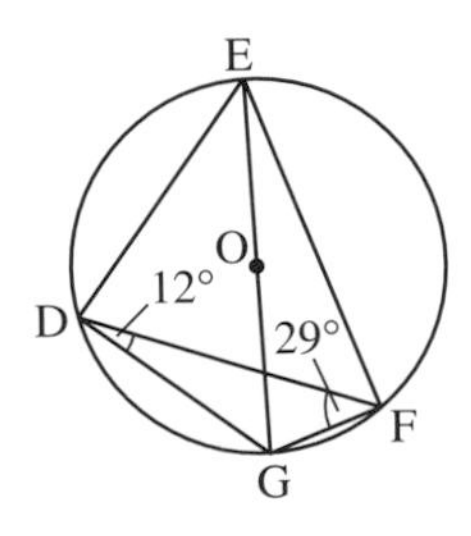

4 Find the values of x and y in each of these circles. O is the centre.

a

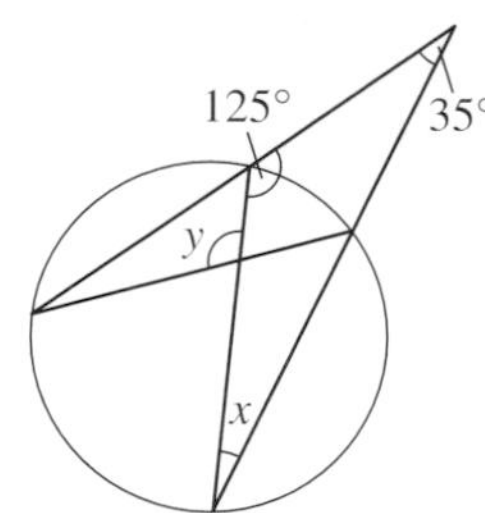

b

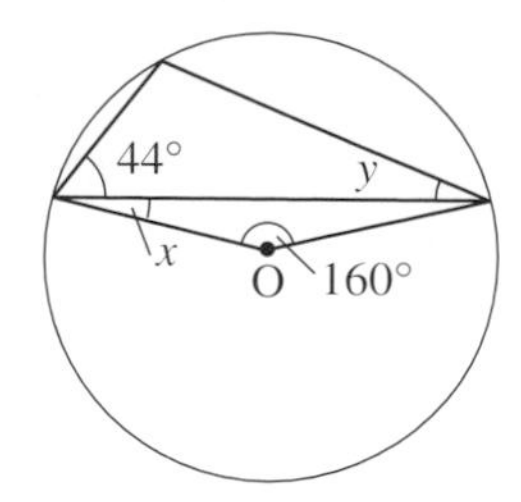

PROOF **5** ABCD are points on a circle. AB is parallel to CD.
Prove that $\angle BAD = \angle ADC$

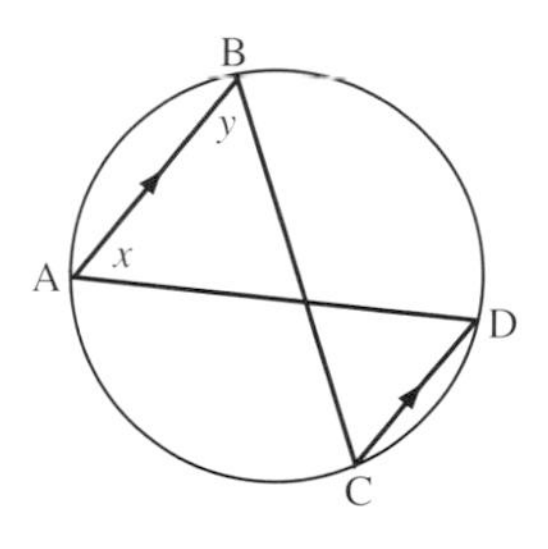

★6 In each diagram, O is the centre of a circle.

a Calculate the value of angle a.

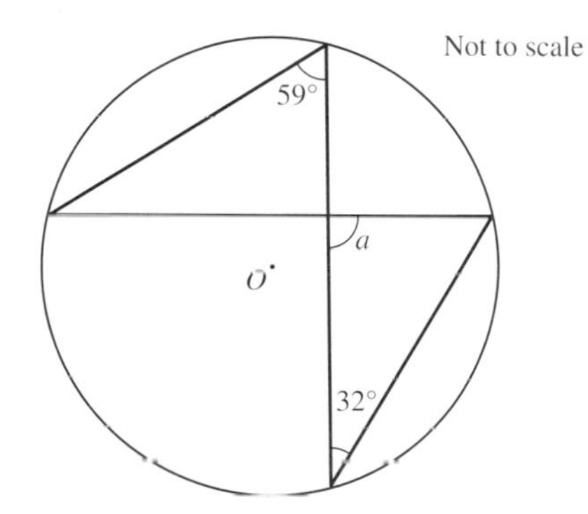

b Calculate the value of angle b.

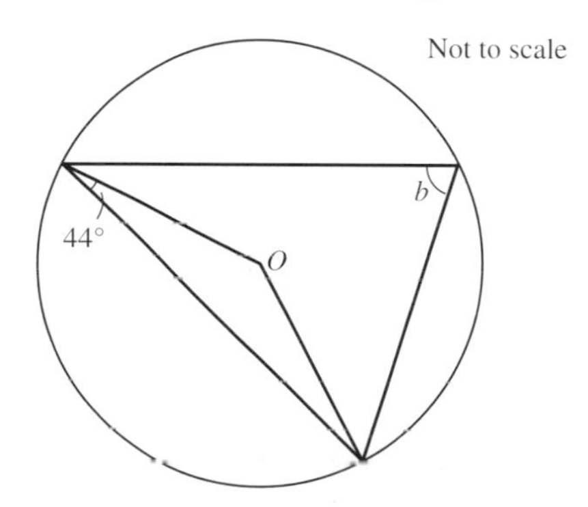

HOMEWORK 5D

1 Find the size of the lettered angles in each of these circles.

a

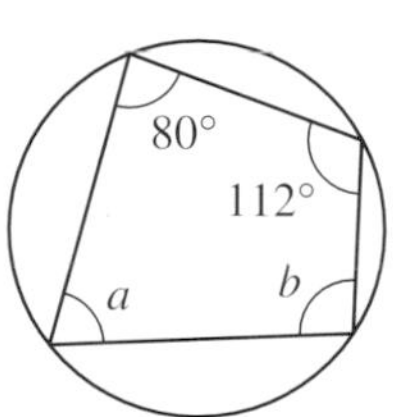

b

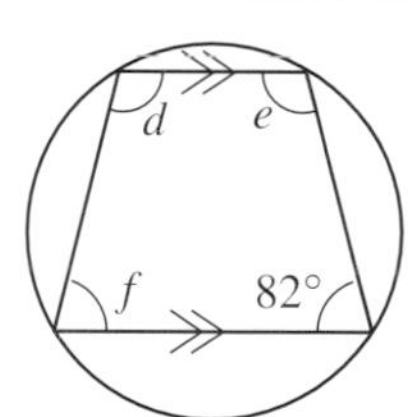

c

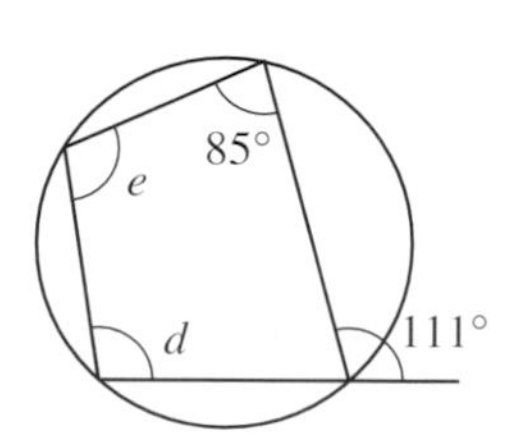

d

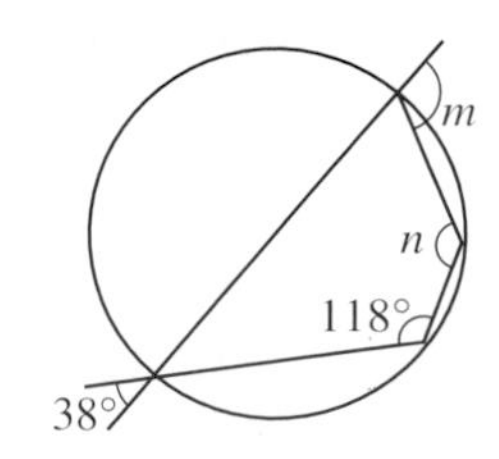

2 Find the values of x and y in each of these circles.

a

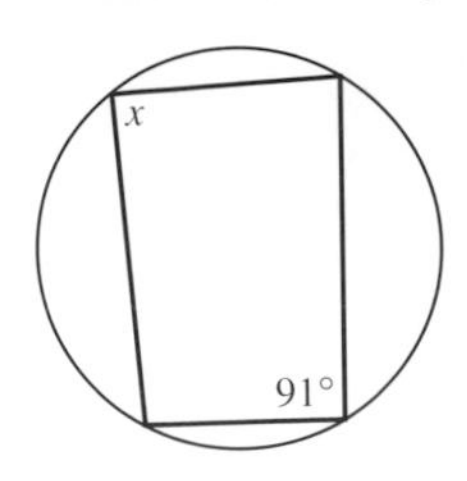

b

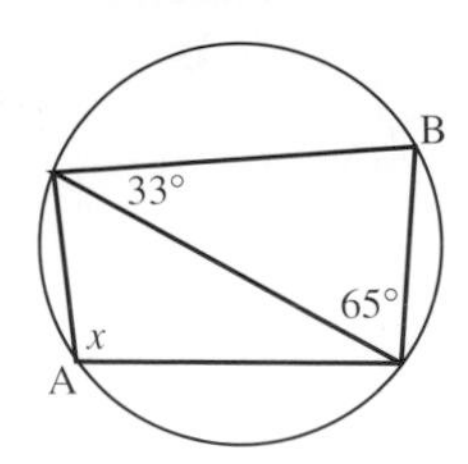

c

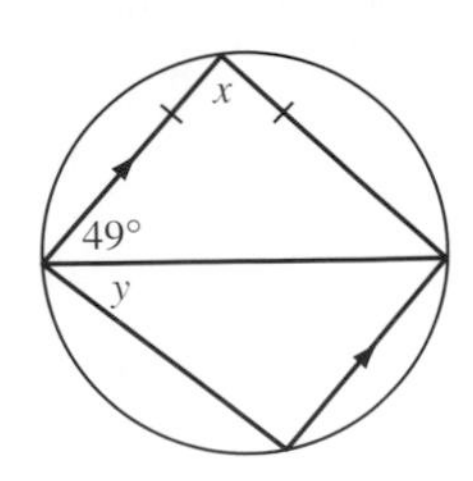

3 Find the values of x and y in each of these circles, centre O.

a

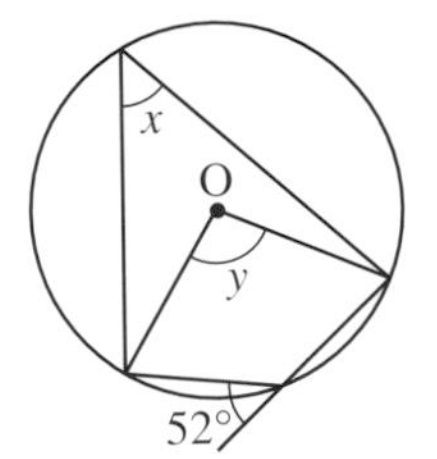

b

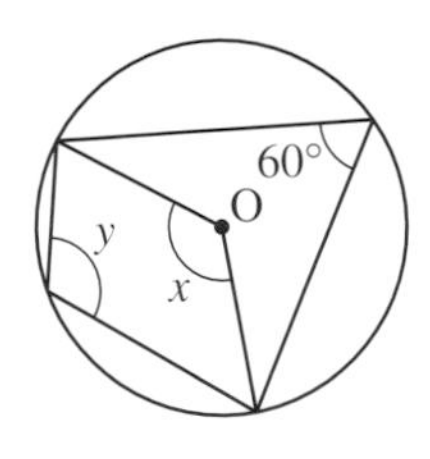

c

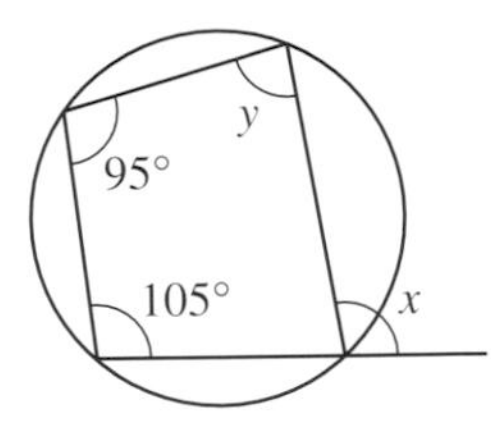

PROOF ★**4** ABCD are points on a circle. AB is parallel to CD.
Prove that ∠BAC = ∠ABD

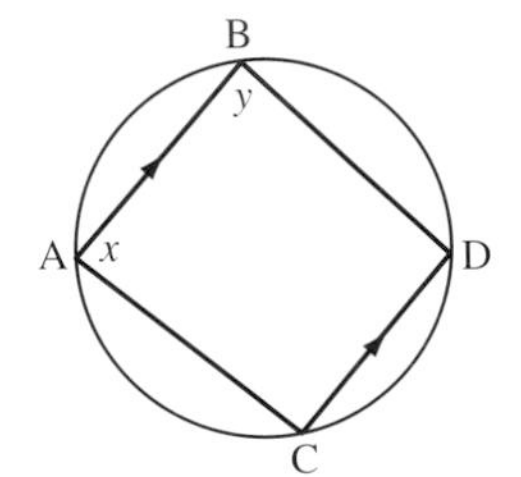

Homework 5E

1 In each diagram, TP and TQ are tangents to a circle, centre O. Find values for r and x.

a

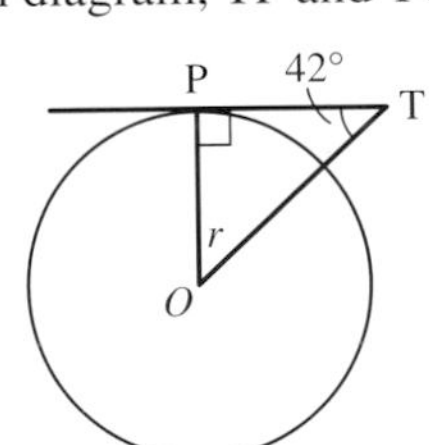

b

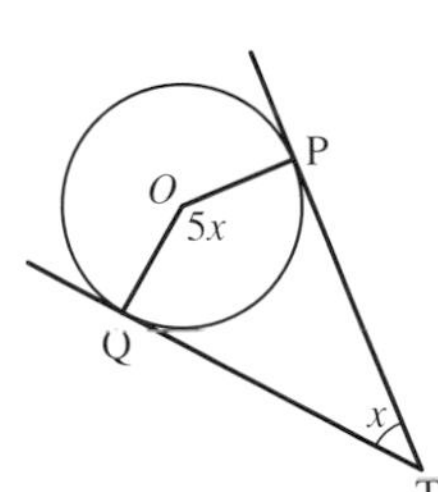

2 Each diagram shows a tangent to a circle, centre O. Find each value of y.

a

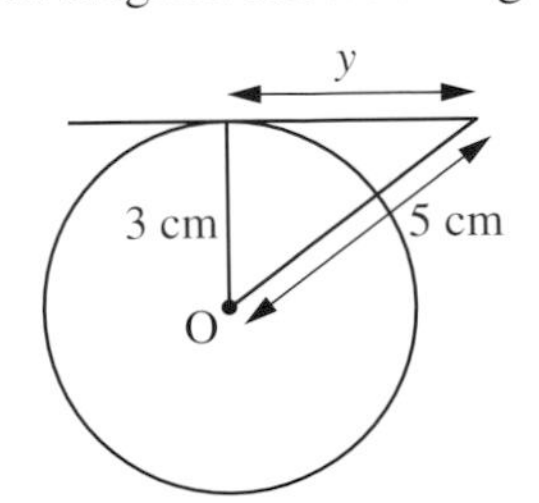

b

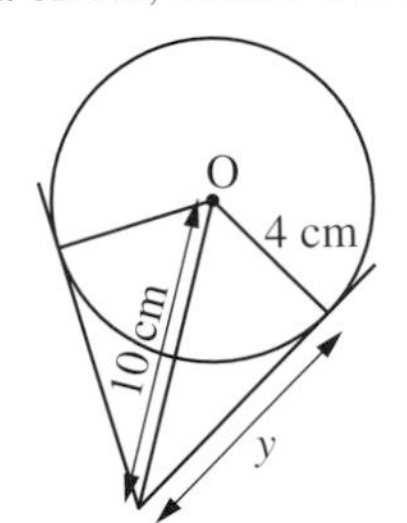

3 Each diagram shows a tangent to a circle, centre O. Find x and y in each case.

a

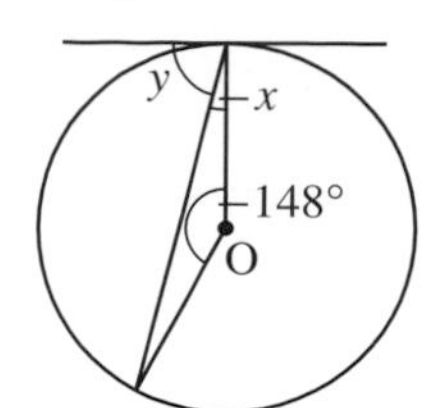

b

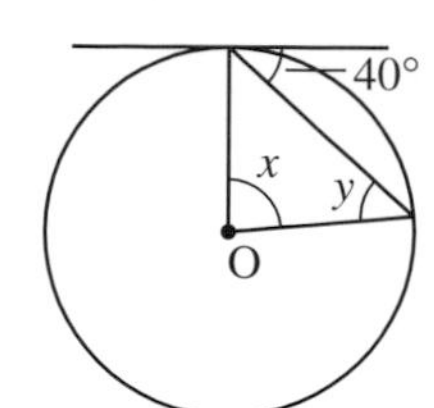

4 In each of the diagrams, TP and TQ are tangents to the circle, centre O. Find each value of x.

a

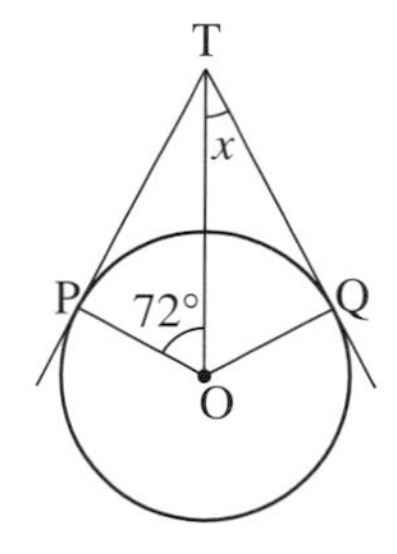

b

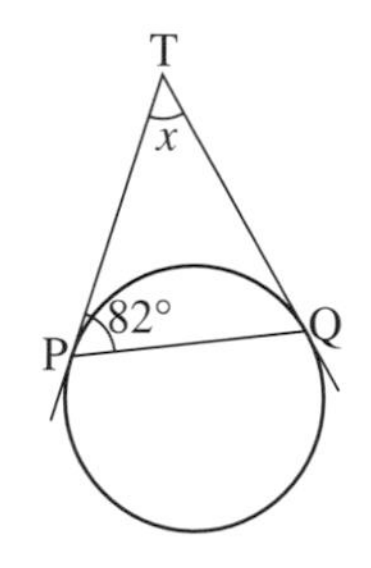

★5 In the diagram, AB = AD and angle ADB = 69°.

a Work out the size of angle BED. Give reasons for each step of your working.

b Work out the size of angle BCD

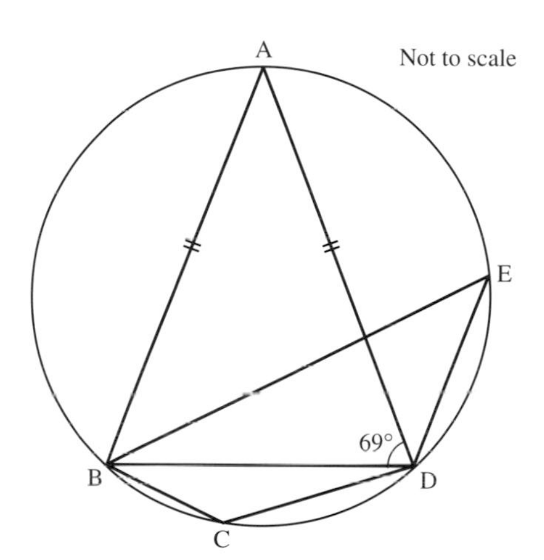

HOMEWORK 5F

1 Find the size of each lettered angle.

a

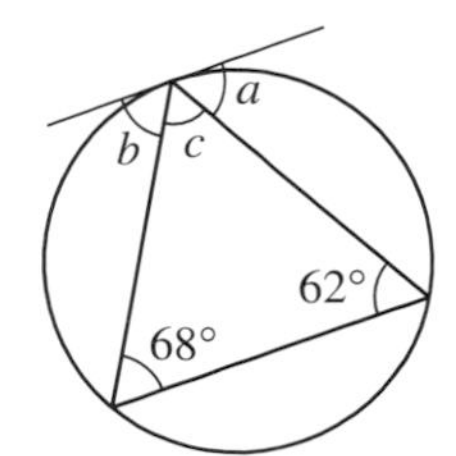

b

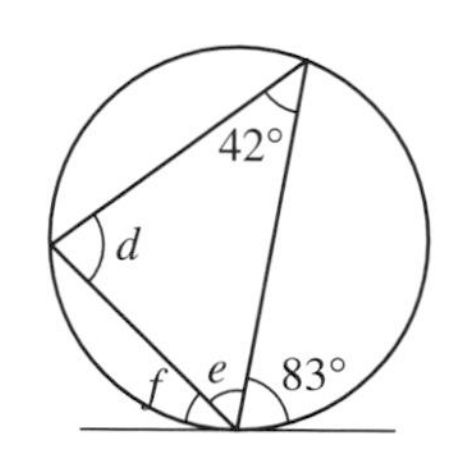

2 In each diagram, find the value of x.

a

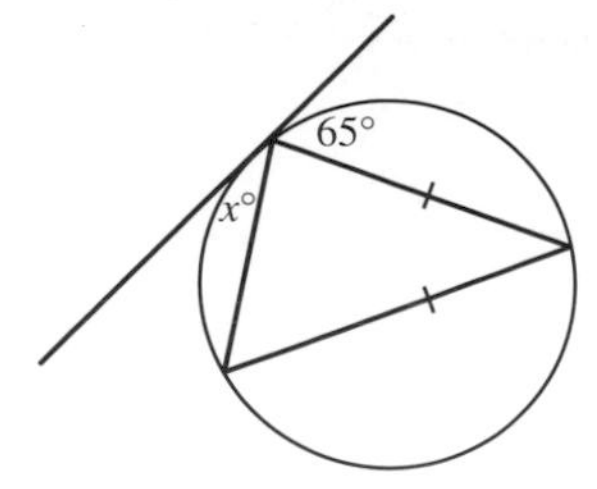

b

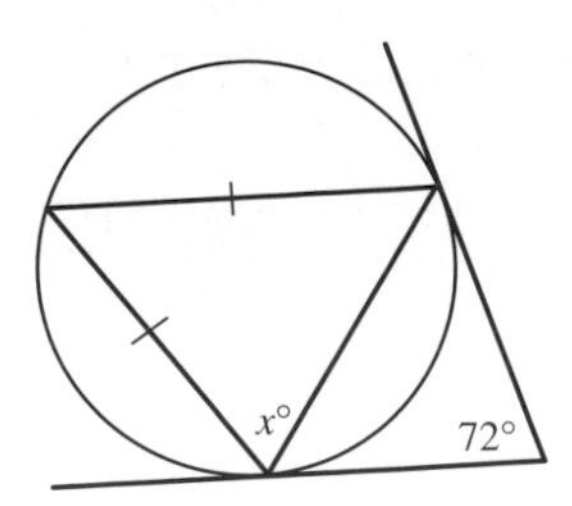

3 In each diagram, find the value of x and y.

a

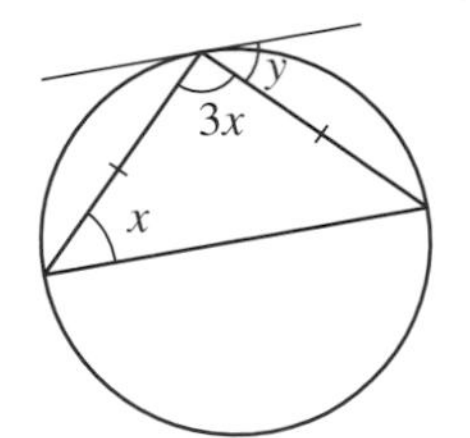

b

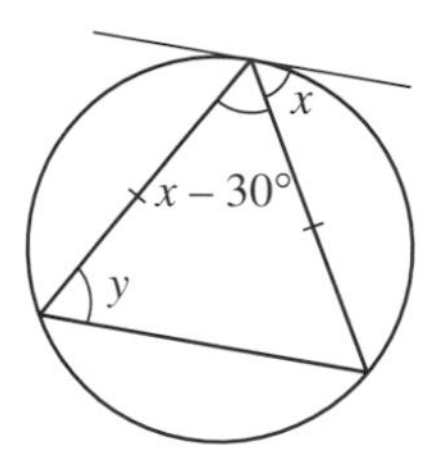

4 ATB is a tangent to the circle, centre O. Find the values of x, y and z in each case.

a

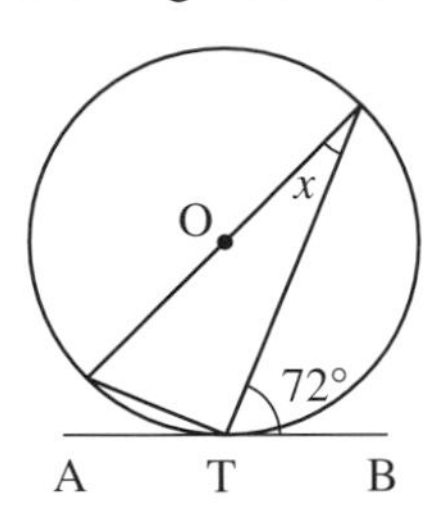

b

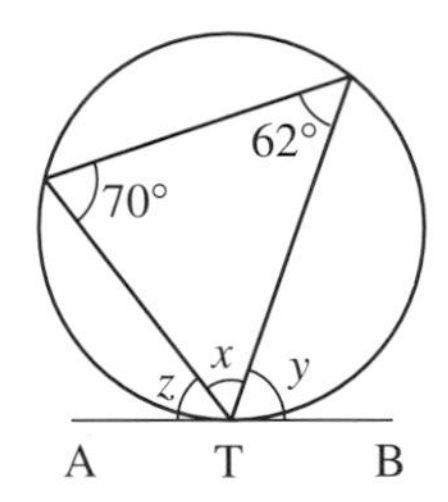

★5 O is the centre of the circle. PQT is the tangent to the circle at Q.
Work out the sizes of angles x, y and z.

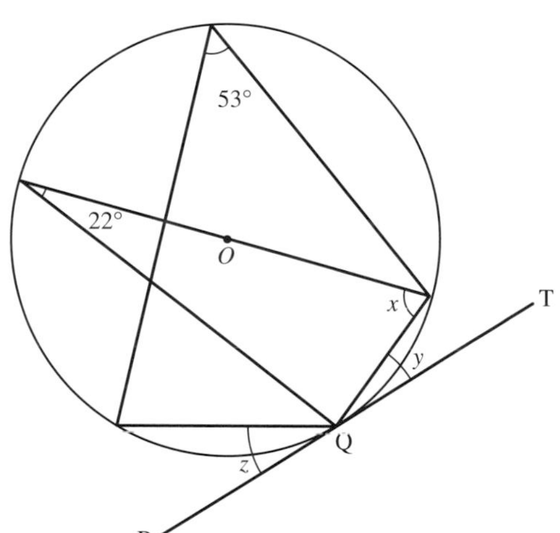

Chapter 6 Transformation geometry

HOMEWORK 6A

1 State whether each pair of triangles below is congruent, giving the reasons if they are.

a

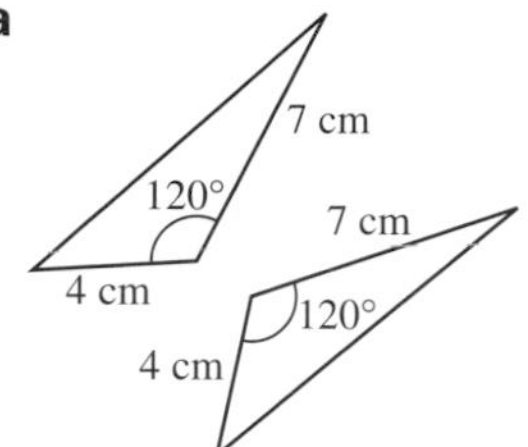

b

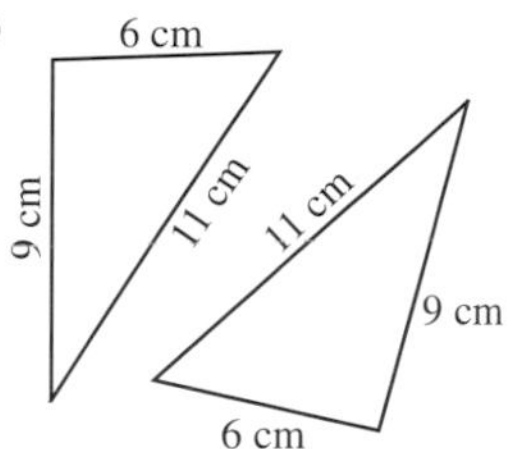

c

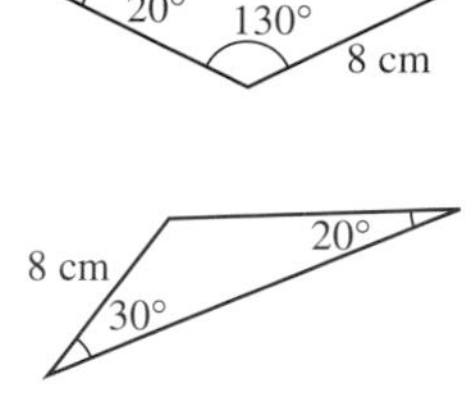

2 Draw a square ABCD. Draw in the diagonals AC and BD. Which triangles are congruent to each other?

3 Draw a kite EFGH. Draw in the diagonals EG and FH. Which triangles are congruent to each other?

4 Draw a rhombus ABCD. Draw in the diagonals AC and BD. Which triangles are congruent to each other?

5 Draw an equilateral triangle ABC. Draw the lines from each vertex to the mid-point of the opposite side. These three lines should all cross at the same point T inside the triangle. Which triangles are congruent to each other?

HOMEWORK 6B

1 Describe with vectors these translations.

i A to B **ii** A to C **iii** A to D **iv** B to A **v** B to C **vi** B to D

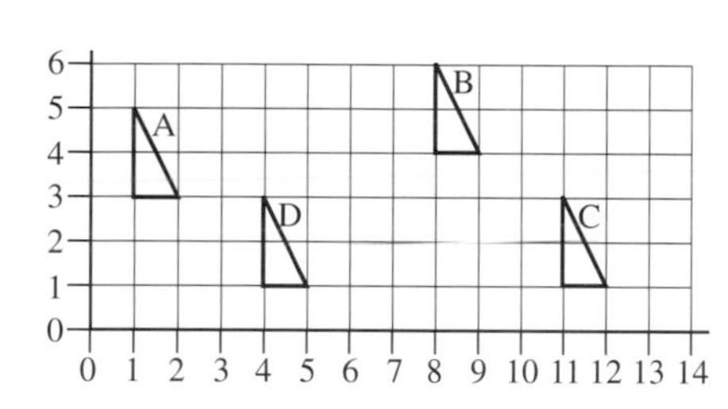

2 **a** On a grid showing values of x and y from 0 to 10, draw the triangle with co-ordinates A(4, 4), B(5, 7) and C(6, 5).

b Draw the image of ABC after a translation with vector $\binom{3}{2}$. Label this P.

c Draw the image of ABC after a translation with vector $\binom{4}{-3}$. Label this Q.

d Draw the image of ABC after a translation with vector $\binom{-4}{3}$. Label this R.

e Draw the image of ABC after a translation with vector $\binom{-3}{-2}$. Label this S.

3 Using your diagram from Question **2**, describe the translation that will move

a P to Q **b** Q to R **c** R to S **d** S to P

e R to P **f** S to Q **g** R to Q **h** P to S

HOMEWORK 6C

1 Copy each shape on squared paper and draw its image after a reflection in the given mirror line.

a

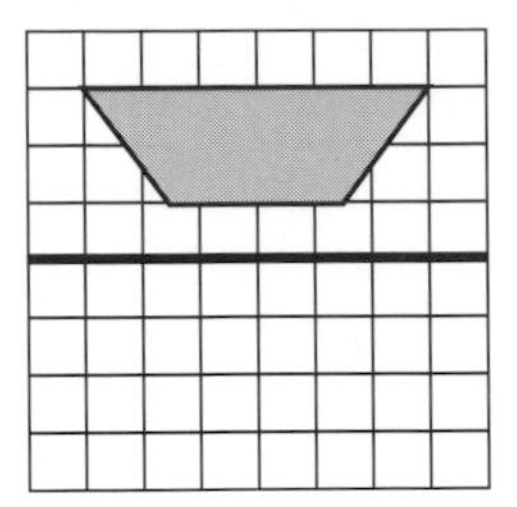

b

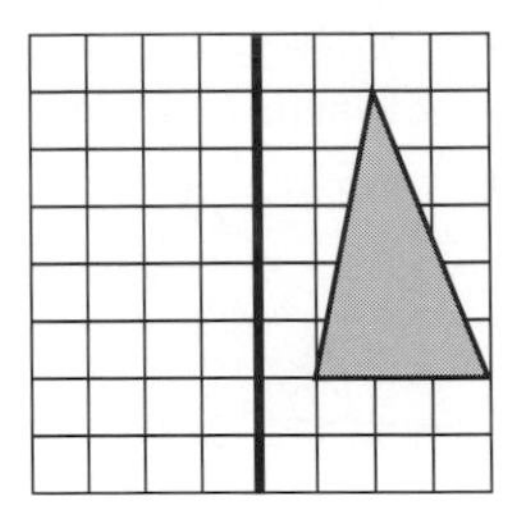

c

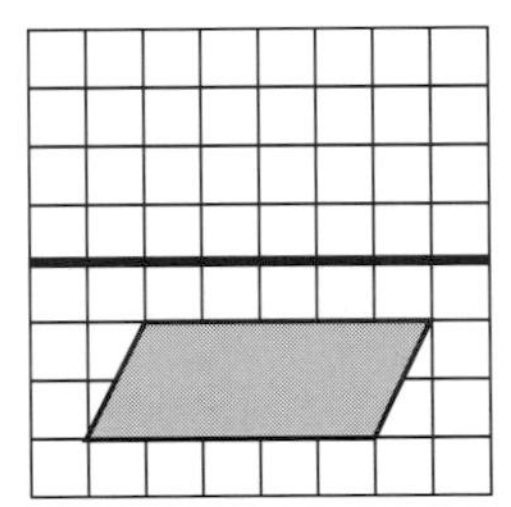

d

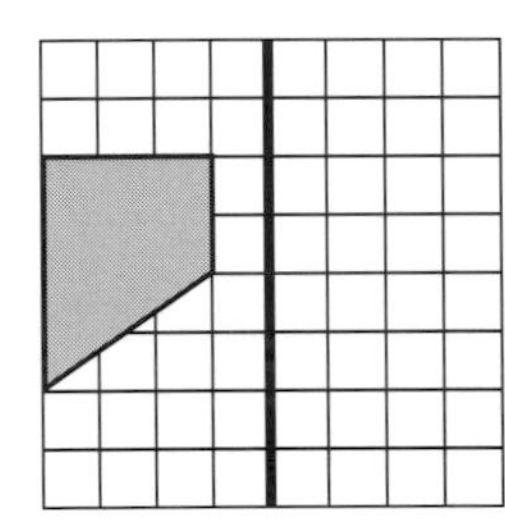

e

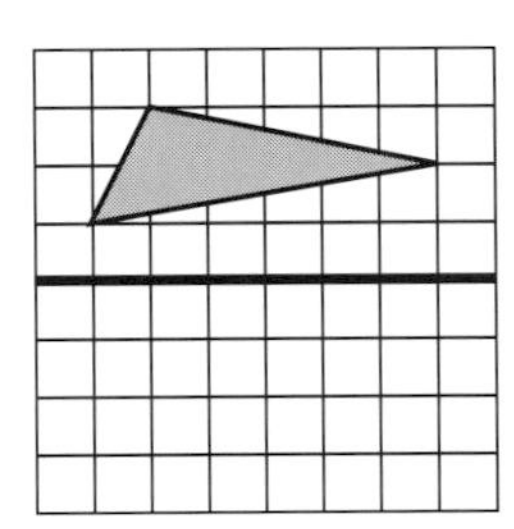

f

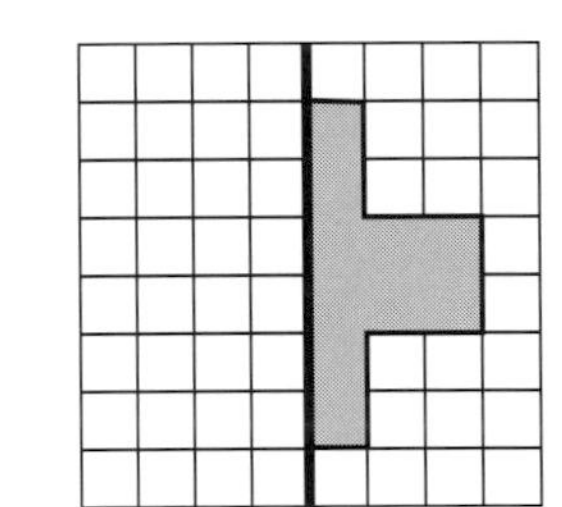

2 **a** Draw a pair of axes, x-axis from –5 to 5, y-axis from –5 to 5.

b Draw the triangle with co-ordinates A(2, 2), B(3, 4), C(2, 4).

c Reflect the triangle ABC in the line $y = x$. Label the image P.

d Reflect the triangle P in the line $y = -x$. Label the image Q.

e Reflect triangle Q in the line $y = x$, label it R.

f Describe the reflection that will move triangle ABC to triangle R.

★**3** Copy this diagram on squared paper.

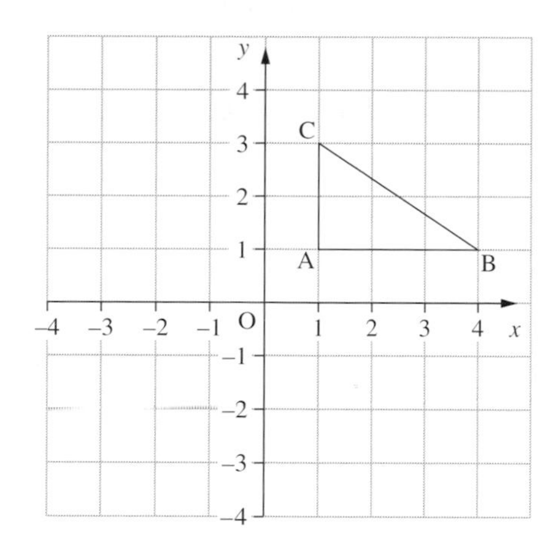

a Reflect the triangle ABC in the x-axis. Label the image R.
b Reflect the triangle ABC in the y-axis. Label the image S.
c What special name is given to figures that are exactly the same shape and size?

4 **a** Draw a pair of axes with the x-axis from –5 to 5 and the y-axis from –5 to 5.
b Draw the triangle with co-ordinates A(1, 1), B(5, 5), C(3, 4).
c Reflect triangle ABC in the x-axis. Label the image P.
d Reflect triangle P in the y-axis. Label the image Q.
e Reflect triangle Q in the x-axis. Label the image R.
f Describe the reflection that will transform triangle ABC onto triangle R.

HOMEWORK 6D

1

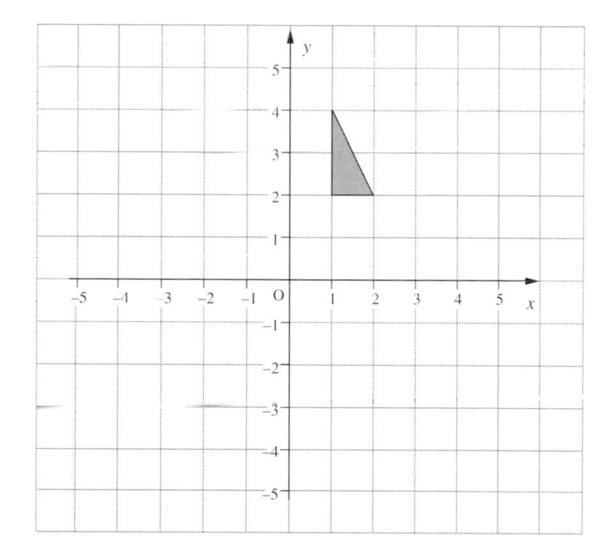

Copy the diagram and rotate the given triangle by
a $\frac{1}{4}$ turn clockwise about (0, 0)
b $\frac{1}{2}$ turn clockwise about (0, 2)
c 90° turn anticlockwise about (–1, 1)
d 180° turn about (0, 0).

2 Describe the rotation that takes the shaded triangle to

a triangle A
b triangle B
c triangle C
d triangle D.

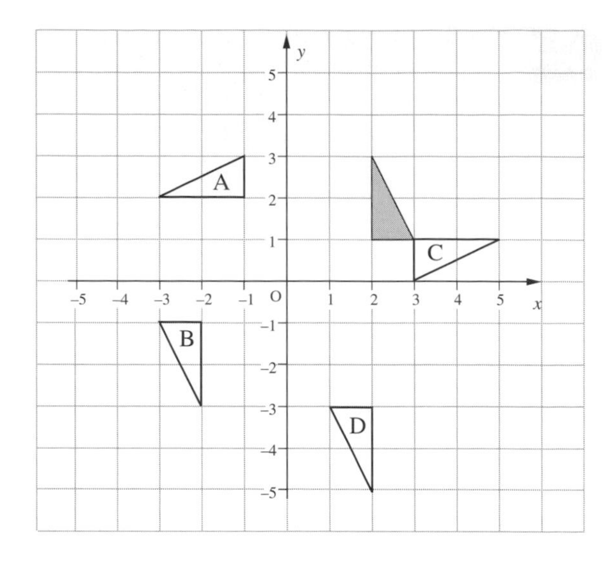

★3 Copy this T-shape on squared paper.

a Rotate the shape 90° clockwise about the origin O. Label the image P.
b Rotate the shape 180° clockwise about the origin O. Label the image Q.
c Rotate the shape 270° clockwise about the origin O. Label the image R.
d What rotation takes R back to the original shape?

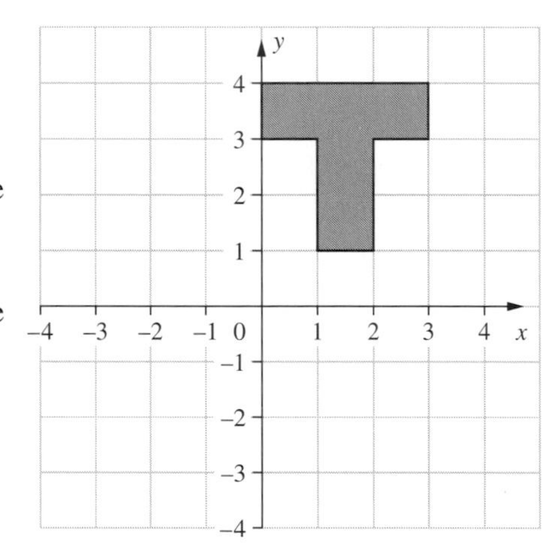

★4 Copy this square ABCD on squared paper.

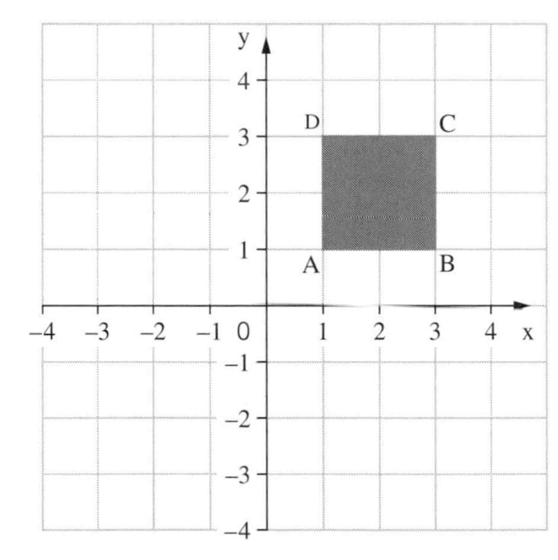

a Write down the co-ordinates of the vertices of the square ABCD.
b Rotate the square ABCD through 90° clockwise about the origin O. Label the image S. Write down the co-ordinates of the vertices of the square S.
c Rotate the square ABCD through 180° clockwise about the origin O. Label the image T. Write down the co-ordinates of the vertices of the square T.
d Rotate the square ABCD through 270° clockwise about the origin O. Label the image U. Write down the co-ordinates of the vertices of the square U.
e What do you notice about the co-ordinates of the four squares?

HOMEWORK 6E

1 Copy each figure below with its centre of enlargement, leaving plenty of space for the enlargement. Then enlarge them by the given scale factor, using the ray method.

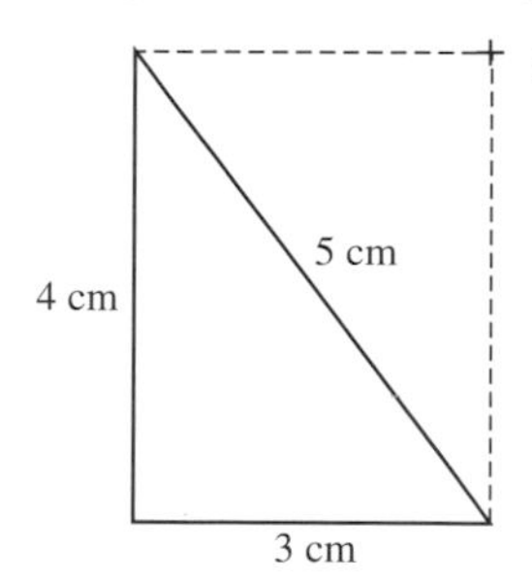

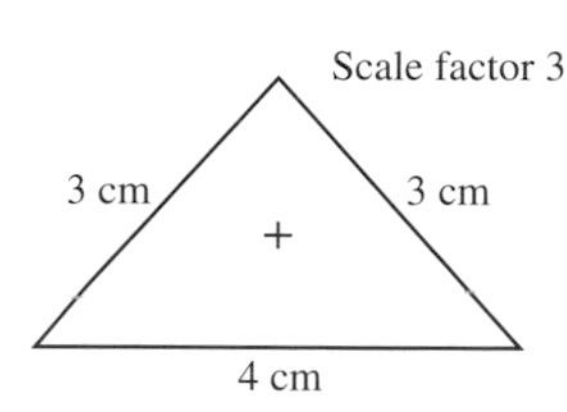

2 Copy each figure below with its centre of enlargement, leaving plenty of space for the enlargement. Then enlarge them by the given scale factor, using the co-ordinate method.

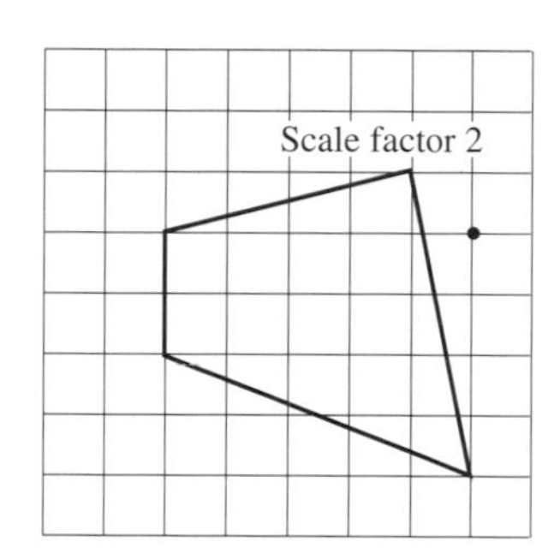

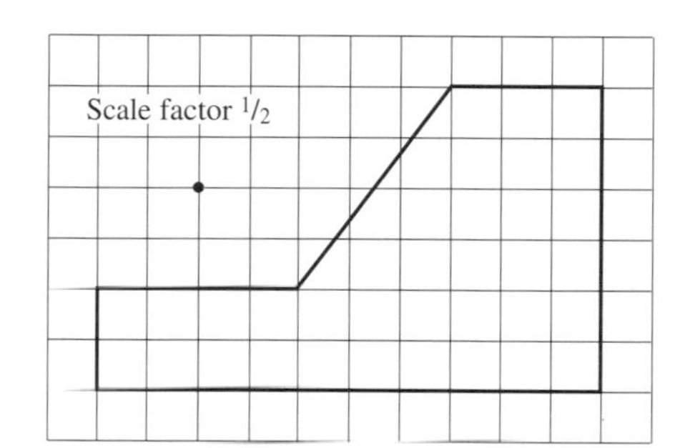

HOMEWORK 6F

1 Describe fully the transformation that will move

a T_1 to T_2 **b** T_1 to T_6 **c** T_2 to T_3 **d** T_6 to T_2

e T_6 to T_5 **f** T_5 to T_4 **g** T_1 to T_5

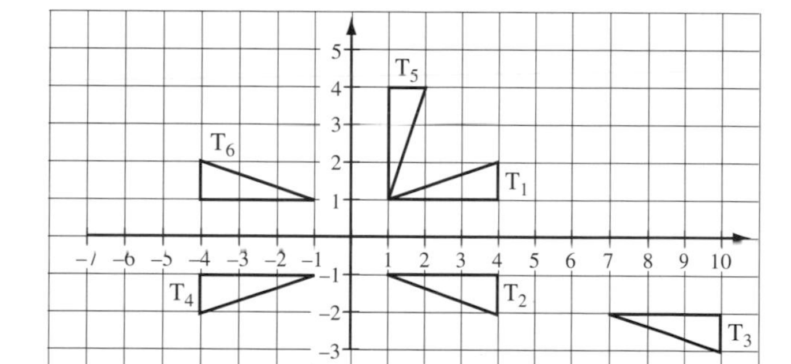

2 **a** Plot a triangle T with vertices (1, 1), (3, 1), (3, 4).

b Reflect triangle T in the x-axis and label the image T_b.

c Rotate triangle T_b 90° clockwise about the origin and label the image T_c.

d Reflect triangle T_c in the x-axis and label the image T_d.

e Describe fully the transformation that will move triangle T_d back to triangle T.

3 The point P(2, 5) is reflected in the x-axis, then rotated by 90° clockwise about the origin. What are the co-ordinates of the image of P?

Chapter 7 Constructions

HOMEWORK 7A

1 Draw a line 8 cm long. Bisect it with a pair of compasses. Check your accuracy by seeing if each half is 4 cm.

2 a Draw any triangle
  b On each side construct the line bisector. All your line bisectors should intersect at the same point.
  c See if you can use this point as the centre of a circle that fits perfectly inside the triangle.

3 a Draw a circle with a radius of about 4 cm.
  b Draw a quadrilateral such that the vertices (corners) of the quadrilateral are on the circumference of the circle.
  c Bisect two of the sides of the quadrilateral. Your bisectors should meet at the centre of the circle.

4 a Draw any angle.
  b Construct the angle bisector.
  c Check how accurate you have been by measuring each half.

★5 The diagram shows a park with two ice-cream sellers A and B. People always go to the ice-cream seller nearest to them. Shade the region of the park from which people go to ice-cream seller B.

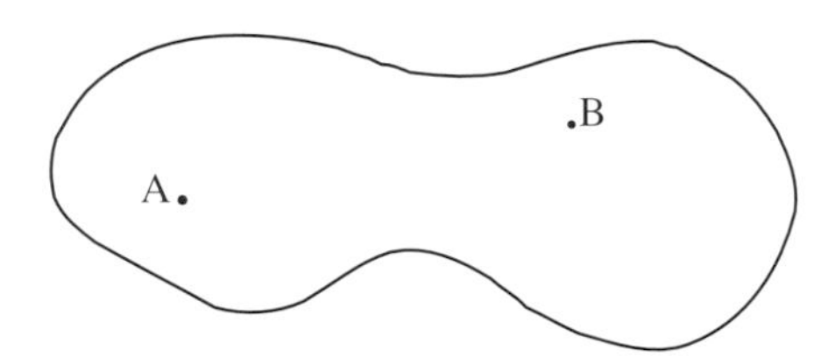

HOMEWORK 7B

1 Construct these triangles accurately without using a protractor.

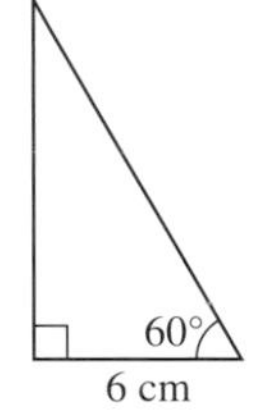

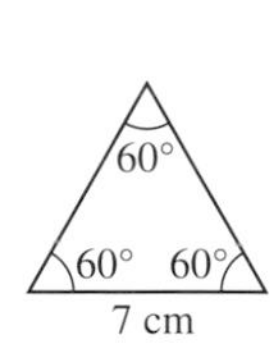

2 a Without using a protractor, construct a square of side 5 cm.
  b See how accurate you have been by constructing an angle bisector on any of the right-angles and seeing whether this also cuts through the opposite right-angle.

3 With ruler and compasses only, construct an angle of 45°.

4 With ruler and compasses only, construct an angle of 30°.

5 a Draw a line 8 cm long. Call it AB.
  b At the point A construct an angle of 60°.
  c At the point B construct an angle of 45°.
  d Label the point where the two lines meet C.
  e Measure AC and BC.

★6 **a** Draw an accurate triangle ABC using only a ruler and compasses.

b Measure the side BC.

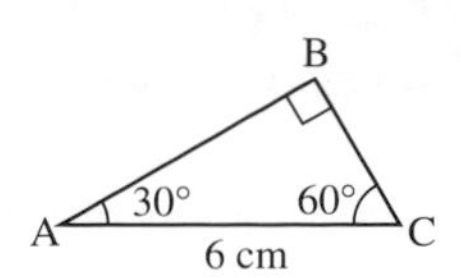

Homework 7C

1 A is a fixed point. Sketch the locus of the point P when AP > 3 cm and AP < 6 cm.

2 A and B are two fixed points 4 cm apart. Sketch the locus of the point P for the following situations:

a AP < BP

b P is always within 3 cm of A and within 2 cm of B.

3 A fly is tethered by a length of spider's web that is 1 m long. Describe the locus that the fly can still buzz about in.

4 ABC is an equilateral triangle of side 4 cm. In each of the following loci, the point P moves only inside the triangle. Sketch the locus in each case.

a AP = BP **b** AP < BP

c CP < 2 cm **d** CP > 3 cm and BP > 3 cm

5 A wheel rolls around the inside of a square. Sketch the locus of the centre of the wheel.

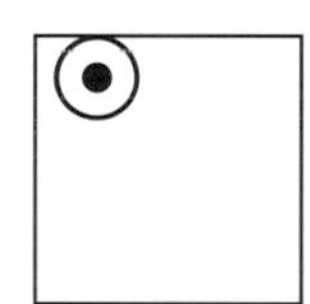

6 The same wheel rolls around the outside of the square. Sketch the locus of the centre of the wheel.

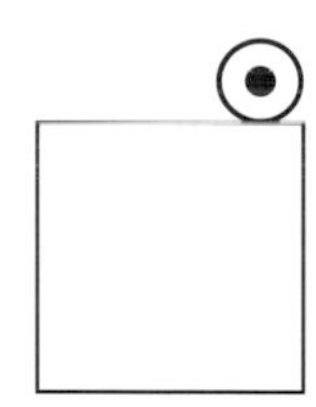

★7 Two ships A and B, which are 7 km apart, both hear a distress signal from a fishing boat. The fishing boat is less than 4 km from ship A and is less than 4.5 km from ship B. A helicopter pilot sees that the fishing boat is nearer to ship A than to ship B. Use accurate construction to show the region which contains the fishing boat. Shade this region.

HOMEWORK 7D

For questions 1 to 3, you should start by sketching the picture given in each question on a 6×6 grid, each square of which is 1 cm by 1 cm. The scale for each question is given.

1 A goat is tethered by a rope, 10 m long, and a stake that is 2 m from each side of a field. What is the locus of the area that the goat can graze? Use a scale of 1 cm : 2 m.

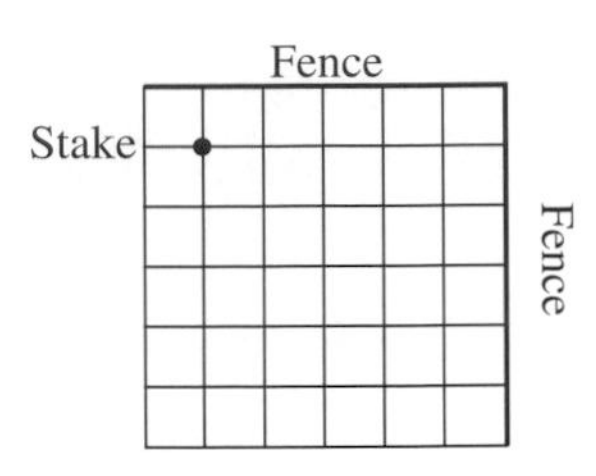

2 A cow is tethered to a rail at the top of a fence 4 m long. The rope is 4 m long. Sketch the area that the cow can graze. Use a scale of 1 cm : 2 m.

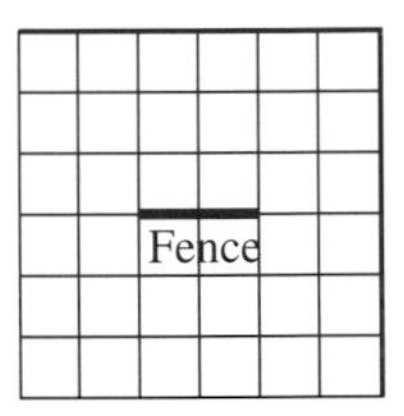

3 A horse is tethered to a corner of a shed, 3 m by 1 m. The rope is 4 m long. Sketch the area that the horse can graze. Use a scale of 1 cm : 1 m.

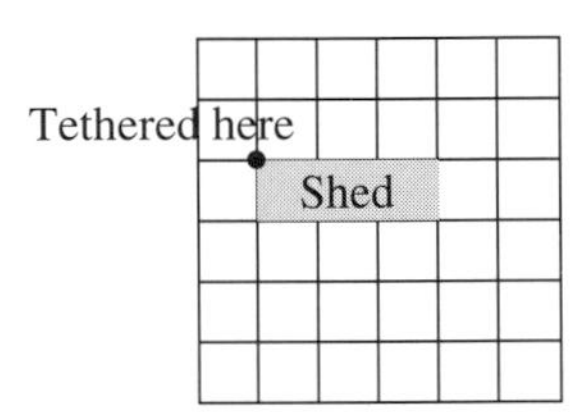

For questions 4 to 6, you should use a copy of the map on page 43. For each question, trace the map and mark on those points that are relevant to that question.

4 A radio station broadcasts from Birmingham with a range that is just far enough to reach York. Another radio station broadcasts from Glasgow with a range that is just far enough to reach Newcastle.

a Sketch the area to which each station can broadcast.

b Will the Birmingham station broadcast as far as Norwich?

c Will they then interfere with each other?

5 An air traffic control centre is to be built in Newcastle. If it has a range of 200 km, will it cover all the area of Britain North of Sheffield and South of Glasgow?

6 A radio transmitter is to be built so that it is the same distance from Exeter, Norwich and Newcastle.

a Draw the perpendicular bisectors of the lines joining these three places to find where the station is to be built.

b Birmingham has so many radio stations that it cannot have another one within 50 km. Can the transmitter be built?

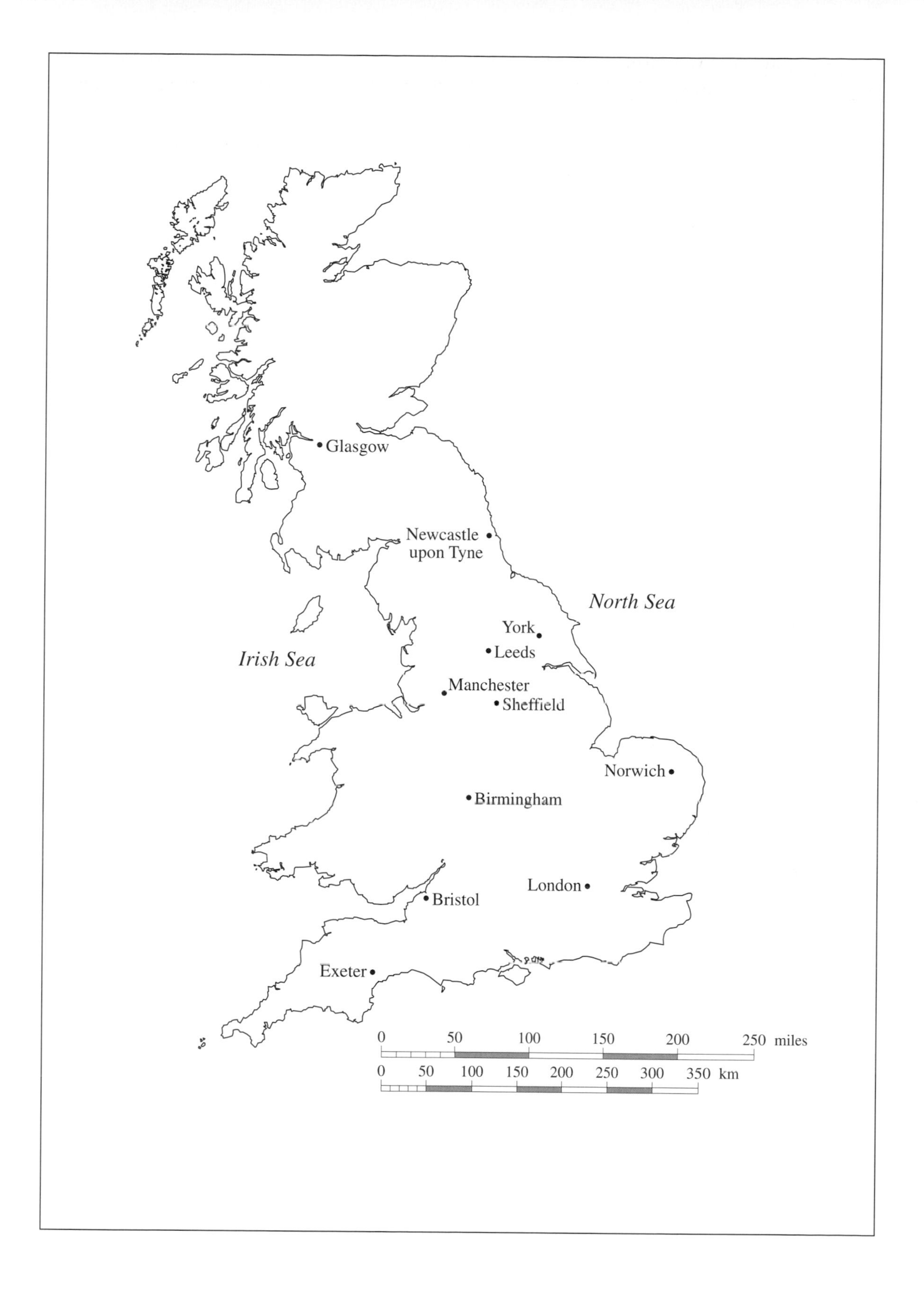
Glasgow
Newcastle upon Tyne
North Sea
York
Leeds
Irish Sea
Manchester
Sheffield
Norwich
Birmingham
London
Bristol
Exeter
0 50 100 150 200 250 miles
0 50 100 150 200 250 300 350 km

★7 A new mobile telephone mast covers the three towns Alphaville, Beeton, and Ceeton. The position of the three towns is shown. The diagram is drawn to a scale of 1 cm to 10 km.

Ceeton

Alphaville

Beeton

The mast is located within the triangle formed by Alphaville, Beeton, and Ceeton so that it is

i equidistant from Alphaville and Ceeton **ii** 50 km from Beeton.

Construct the position of the mast on the diagram. Mark with an X the position of the mast.

Chapter 8 Number 2

Homework 8A

1 Write these expressions using power notation. Do not work them out yet.

a $5 \times 5 \times 5 \times 5$
b $7 \times 7 \times 7 \times 7 \times 7$
c $19 \times 19 \times 19$
d $4 \times 4 \times 4 \times 4 \times 4$
e $1 \times 1 \times 1 \times 1 \times 1 \times 1 \times 1$
f $8 \times 8 \times 8 \times 8 \times 8$
g 6
h $11 \times 11 \times 11 \times 11 \times 11 \times 11$
i $0.9 \times 0.9 \times 0.9 \times 0.9$
j $999 \times 999 \times 999$

2 Write these power terms out in full. Do not work them out yet.

a 4^5 **b** 8^4 **c** 5^3 **d** 9^6 **e** 1^{11}
f 7^3 **g** 5.2^3 **h** 7.5^3 **i** 7.7^4 **j** $10\,000^3$

3 Using the power key on your calculator (or another method), work out the values of the power terms in Question **1**.

4 Using the power key on your calculator (or another method), work out the values of the power terms in Question **2**.

5 Without using a calculator, work out the values of these power terms.

a 7^0 **b** 9^1 **c** 17^0 **d** 1^{91} **e** 10^5

6 Using your calculator, or otherwise, work out the values of these power terms.

a $(-2)^3$ **b** $(-1)^{11}$ **c** $(-3)^4$ **d** $(-5)^3$ **e** $(-10)^6$

7 Without using a calculator, write down the answers to these.

a $(-4)^2$ **b** $(-5)^3$ **c** $(-3)^4$ **d** $(-2)^5$ **e** $(-1)^6$

HOMEWORK 8B

1 Write down each of these in fraction form.

a 5^{-2} **b** 4^{-1} **c** 10^{-3} **d** 3^{-3} **e** x^{-2} **f** $5t^{-1}$

2 Write down each of these in negative index form.

a $\frac{1}{2^4}$ **b** $\frac{1}{7}$ **c** $\frac{1}{x^2}$

3 Change each of the following expressions into an index form of the type shown.

a	All of the form 2^n	**i**	32	**ii**	$\frac{1}{4}$
b	All of the form 10^n	**i**	10 000	**ii**	$\frac{1}{100}$
c	All of the form 5^n	**i**	625	**ii**	$\frac{1}{125}$

4 Find the value of each of the following, where the letters have the given values.

a	Where $x = 3$	**i**	x^2	**ii**	$4x^{-1}$
b	Where $t = 5$	**i**	t^{-2}	**ii**	$5t^{-4}$
c	Where $m = 2$	**i**	m^{-3}	**ii**	$4m^{-2}$

★5 $a = 3$ and $b = 2$. Calculate the value of

a $3a^{-1} + 2b^{-2}$, giving your answer as a fraction in its simplest form.

b $a^{-2} + b^{-3}$, giving your answer as a fraction in its simplest form.

HOMEWORK 8C

1 Write these as single powers of 7.

a $7^3 \times 7^2$ **b** $7^3 \times 7^6$ **c** $7^4 \times 7^3$ **d** 7×7^5 **e** $7^5 \times 7^9$ **f** 7×7^7

2 Write these as single powers of 5.

a $5^6 \div 5^2$ **b** $5^8 \div 5^2$ **c** $5^4 \div 5^3$ **d** $5^5 \div 5^5$ **e** $5^6 \div 5^4$

3 Simplify these and write them as single powers of a.

a $a^2 \times a$ **b** $a^3 \times a^2$ **c** $a^4 \times a^3$ **d** $a^6 \div a^2$ **e** $a^3 \div a$ **f** $a^5 \div a^4$

4 Simplify these expressions.

a $3a^4 \times 5a^2$ **b** $3a^4 \times 7a$ **c** $5a^4 \times 6a^2$ **d** $3a^2 \times 4a^7$
e $5a^4 \times 5a^2 \times 5a^2$

5 Simplify these expressions.

a $8a^5 \div 2a^2$ **b** $12a^7 \div 4a^2$ **c** $25a^6 \div 5a$ **d** $48a^8 \div 6a^{-1}$
e $24a^6 \div 8a^{-2}$ **f** $36a \div 6a^5$

6 Simplify these expressions.

a $3a^3b^2 \times 4a^3b$ **b** $7a^3b^5 \times 2ab^3$ **c** $4a^3b^5 \times 5a^4b^{-1}$

d $12a^3b^5 \div 4ab$ **e** $24a^3b^5 \div 6a^2b^{-3}$

! PROOF 7 Use the general rule for dividing powers of the same number

$$a^x \div a^y = a^{x-y}$$

to prove that any number raised to the power –1 is the reciprocal of that number.

HOMEWORK 8D

Evaluate the following.

1 $36^{\frac{1}{2}}$ **2** $144^{\frac{1}{2}}$ **3** $25^{\frac{1}{2}}$ **4** $196^{\frac{1}{2}}$ **5** $8^{\frac{1}{3}}$ **6** $125^{\frac{1}{3}}$

7 $32^{-\frac{1}{5}}$ **8** $144^{-\frac{1}{2}}$ **9** $27^{-\frac{1}{3}}$ **10** $(\frac{25}{81})^{\frac{1}{2}}$ **11** $(\frac{81}{36})^{\frac{1}{2}}$ **12** $(\frac{36}{64})^{\frac{1}{2}}$

13 $(\frac{8}{27})^{\frac{1}{3}}$ **14** $(\frac{16}{625})^{\frac{1}{4}}$ **15** $(\frac{4}{9})^{-\frac{1}{2}}$ **16** $(\frac{16}{25})^{-\frac{1}{2}}$ **17** $(\frac{8}{27})^{-\frac{1}{3}}$

★18 Evaluate

a $81^{\frac{1}{2}}$ **b** $64^{\frac{1}{6}}$ **c** $64^{-\frac{1}{3}}$

HOMEWORK 8E

1 Evaluate the following.

a $16^{\frac{3}{4}}$ **b** $125^{\frac{4}{3}}$ **c** $81^{\frac{3}{4}}$

2 Rewrite the following in index form.

a $\sqrt[4]{t^3}$ **b** $\sqrt[5]{m^2}$

3 Evaluate the following.

a $27^{\frac{2}{3}}$ **b** $8^{\frac{4}{3}}$ **c** $36^{\frac{3}{2}}$ **d** $81^{1.25}$

4 **a** Draw the graph of $y = 2^x$ for $x = -2$ to 5.

b Use your graph to estimate the value of $2^{\frac{5}{2}}$.

5 Using a trial-and-improvement method, or otherwise, solve these equations.

a $6^x = 60$ **b** $10^x = 2$ (You could try to use the **power** key on your calculator.)

★6 **a** Evaluate $8^{\frac{1}{3}}$. **b** Write $16^{-\frac{1}{2}} \times 2^{-3}$ as a power of 2.

c Given that $32^y = 2$, find the value of y.

HOMEWORK 8F

1 Evaluate the following.

a 3.5×100 **b** 2.15×10 **c** 6.74×1000 **d** 4.63×10

e 30.145×10 **f** 78.56×1000 **g** 6.42×10^2 **h** 0.067×10

i 0.085×10^3 **j** 0.798×10^5 **k** 0.658×1000 **l** 215.3×10^2

m 0.889×10^6 **n** 352.147×10^2 **o** 37.2841×10^3 **p** 34.28×10^6

2 Evaluate the following.

a $4538 \div 100$ **b** $435 \div 10$ **c** $76459 \div 1000$ **d** $643.7 \div 10$

e $4228.7 \div 100$ **f** $278.4 \div 1000$ **g** $246.5 \div 10^2$ **h** $76.3 \div 10$

i $76 \div 10^3$ **j** $897 \div 10^5$ **k** $86.5 \div 1000$ **l** $1.5 \div 10^2$

m $0.8799 \div 10^6$ **n** $23.4 \div 10^2$ **o** $7654 \div 10^3$ **p** $73.2 \div 10^6$

3 Evaluate the following.

a 400×300 **b** 50×4000 **c** 70×200 **d** 30×700
e $(30)^2$ **f** $(50)^3$ **g** $(200)^2$ **h** 40×150
i 70×200 **j** 60×5000 **k** 30×250 **l** 700×200

4 Evaluate the following.

a $4000 \div 800$ **b** $9000 \div 30$ **c** $7000 \div 200$ **d** $8000 \div 200$
e $2100 \div 700$ **f** $9000 \div 60$ **g** $700 \div 50$ **h** $3500 \div 70$
i $3000 \div 500$ **j** $30\,000 \div 2000$ **k** $5600 \div 1400$ **l** $6000 \div 30$

5 Evaluate the following.

a 7.3×10^2 **b** 3.29×10^5 **c** 7.94×10^3 **d** 6.8×10^7
e $3.46 \div 10^2$ **f** $5.07 \div 10^4$ **g** $2.3 \div 10^4$ **h** $0.89 \div 10^3$

Homework 8G

1 Write these standard form numbers out in full.

a 3.5×10^2 **b** 4.15×10 **c** 5.7×10^{-3} **d** 1.46×10
e 3.89×10^{-2} **f** 4.6×10^3 **g** 2.7×10^2 **h** 8.6×10
i 4.6×10^3 **j** 3.97×10^5 **k** 3.65×10^{-3} **l** 7.05×10^2

2 Write these numbers in standard form.

a 780 **b** 0.435 **c** 67 800 **d** 7 400 000 000
e 30 780 000 000 **f** 0.000 427 8 **g** 6450 **h** 0.047
i 0.000 12 **j** 96.43 **k** 74.78 **l** 0.004 157 8

In Questions **3** to **7**, write the appropriate numbers given in each question in standard form.

3 In 1990 there were 24 673 000 vehicles licenced in the UK.

4 In 2001 Keith Gordon was one of 15282 runners to complete the Boston Marathon.

5 In 1990 the total number of passenger kilometres on the British roads was 613 000 000 000.

6 The Sun is 93 million miles away from Earth. The next nearest star to the Earth is Proxima Centuri which is about 24 million million miles away.

7 A Scientist was working with a new particle reported to weigh only 0.000 000 000 000 65 g.

Homework 8H

1 Find the results of the following, leaving your answers in standard form.

a $(4 \times 10^6) \times (7 \times 10^9)$ **b** $(7 \times 10^5) \times (5 \times 10^7)$
c $(3 \times 10^{-5}) \times (8 \times 10^8)$ **d** $(2.1 \times 10^7) \times (5 \times 10^{-8})$

2 Find the results of the following, leaving your answers in standard form.

a $(9 \times 10^8) \div (3 \times 10^4)$ **b** $(2.7 \times 10^7) \div (9 \times 10^3)$
c $(5.5 \times 10^4) \div (1.1 \times 10^{-2})$ **d** $(4.2 \times 10^{-9}) \div (3 \times 10^{-8})$

3 Find the results of the following, leaving your answers in standard form.

a $\dfrac{8 \times 10^9}{4 \times 10^7}$ **b** $\dfrac{12 \times 10^6}{3 \times 10^4}$ **c** $\dfrac{2.8 \times 10^7}{7 \times 10^{-4}}$

4 $p = 8 \times 10^5$ and $q = 2 \times 10^7$

Find the value of the following, leaving your answer in standard form.

a $p \times q$ **b** $p \div q$ **c** $p + q$ **d** $q - p$ **e** $\dfrac{q}{p}$

5 $p = 2 \times 10^{-2}$ and $q = 4 \times 10^{-3}$

a $p \times q$ **b** $p \div q$ **c** $p + q$ **d** $q - p$ **e** $\dfrac{q}{p}$

HOMEWORK 8I

1 State which of the following are irrational.

a π **b** $\sqrt{9}$ **c** $\sqrt{0.64}$ **d** $\pi + 2$ **e** $1\frac{2}{3}$

2 Write down an irrational number between

a 3 and 4 **b** 5 and 6

3 What can be added to each of the following to make a rational number?

a $2 + \sqrt{7}$ **b** $2 - \pi$ **c** $\sqrt{3} - 4$

4 What can be multiplied to each of the following to make a rational number?

a $\sqrt{8}$ **b** $\dfrac{1}{\pi}$

HOMEWORK 8J

1 Work out each of these fractions as a decimal. Give them as terminating decimals or recurring decimals as appropriate.

a $\frac{3}{4}$ **b** $\frac{1}{15}$ **c** $\frac{1}{25}$ **d** $\frac{1}{11}$ **e** $\frac{1}{20}$

2 There are several patterns to be found in recurring decimals. For example,

$\frac{1}{13} = 0.076923076923076923076923\ldots$; $\frac{2}{13} = 0.153846153846153846153846\ldots$

$\frac{3}{13} = 0.230769230769230769230769\ldots$ and so on

a Write down the decimals for $\frac{4}{13}, \frac{5}{13}, \frac{6}{13}, \frac{7}{13}, \frac{8}{13}, \frac{9}{13}, \frac{10}{13}, \frac{11}{13}, \frac{12}{13}$ to 24 decimal places.

b What do you notice?

3 Write each of these fractions as a decimal. Use this to write the list in order of size, smallest first.

$\frac{2}{9}$ $\frac{1}{5}$ $\frac{23}{100}$ $\frac{2}{7}$ $\frac{3}{11}$

4 Convert each of these terminating decimals to a fraction.

a 0.57 **b** 0.275 **c** 0.85 **d** 0.06 **e** 3.65

5 $x = 0.0242\,424\ldots$

a What is $100x$?

b By subtracting the original value from your answer to part **a**, work out the value of $99x$.

c Multiply both sides by 10 to get $990x$ and eliminate the decimal on the right-hand side.

d Divide both sides by 990.

e What is x as a fraction expressed in its lowest terms?

6 Convert each of these recurring decimals to a fraction.

a $0.\dot{7}$ **b** $0.\dot{5}\dot{7}$ **c** $0.\dot{5}\dot{4}$ **d** $0.\dot{2}7\dot{5}$

i $2.\dot{5}$ **j** $2.\dot{3}\dot{6}$ **k** $0.0\dot{6}\dot{3}$ **l** $2.0\dot{7}\dot{5}$

★7 **a** Write 1.7 as a rational number in the form $\frac{a}{b}$, where a and b are whole numbers.

Given that $n = 1.\dot{7}$

b **i** Write down the value of $10n$.

ii Hence write down the value of $9n$.

iii Express n as a rational number, in the form $\frac{a}{b}$, where a and b are whole numbers.

HOMEWORK 8K

1 Work out each of the following in simplified form.

a $\sqrt{3} \times \sqrt{4}$ **b** $\sqrt{5} \times \sqrt{7}$ **c** $\sqrt{5} \times \sqrt{5}$ **d** $\sqrt{2} \times \sqrt{32}$

2 Work out each of the following in surd form.

a $\sqrt{15} \div \sqrt{5}$ **b** $\sqrt{18} \div \sqrt{2}$ **c** $\sqrt{32} \div \sqrt{2}$ **d** $\sqrt{12} \div \sqrt{8}$

3 Work out each of the following in surd form.

a $\sqrt{3} \times \sqrt{3} \times \sqrt{2}$ **b** $\sqrt{5} \times \sqrt{5} \times \sqrt{15}$ **c** $\sqrt{2} \times \sqrt{8} \times \sqrt{8}$ **d** $\sqrt{2} \times \sqrt{8} \times \sqrt{5}$

4 Work out each of the following in surd form.

a $\sqrt{3} \times \sqrt{8} \div \sqrt{2}$ **b** $\sqrt{15} \times \sqrt{3} \div \sqrt{5}$ **c** $\sqrt{8} \times \sqrt{8} \div \sqrt{2}$ **d** $\sqrt{3} \times \sqrt{27} \div \sqrt{3}$

5 Simplify each of the following surds into the form $a\sqrt{b}$.

a $\sqrt{90}$ **b** $\sqrt{32}$ **c** $\sqrt{63}$ **d** $\sqrt{300}$

e $\sqrt{150}$ **f** $\sqrt{270}$ **g** $\sqrt{96}$ **h** $\sqrt{125}$

6 Simplify each of these.

a $2\sqrt{32} \times 5\sqrt{2}$ **b** $4\sqrt{8} \times 2\sqrt{2}$ **c** $4\sqrt{12} \times 5\sqrt{3}$ **d** $3\sqrt{6} \times 2\sqrt{6}$

e $2\sqrt{5} \times 5\sqrt{3}$ **f** $2\sqrt{3} \times 3\sqrt{3}$ **g** $2\sqrt{2} \times 3\sqrt{8}$ **h** $2\sqrt{3} \times 2\sqrt{27}$

i $8\sqrt{24} \div 2\sqrt{3}$ **j** $3\sqrt{27} \div \sqrt{3}$ **k** $5\sqrt{18} \div \sqrt{2}$ **l** $2\sqrt{32} \div 4\sqrt{8}$

m $5\sqrt{2} \times \sqrt{8} \div 2\sqrt{2}$ **n** $3\sqrt{15} \times \sqrt{3} \div \sqrt{5}$ **o** $2\sqrt{24} \times 5\sqrt{3} \div 2\sqrt{8}$

7 Find the value of a that makes each of these surds true.

a $\sqrt{5} \times \sqrt{a} = 20$ **b** $\sqrt{3} \times \sqrt{a} = 12$ **c** $\sqrt{5} \times 4\sqrt{a} = 20$

! PROOF **8** The following rule is not true. Try a numerical example to show this.

$\sqrt{(a^2 + b^2)} = a + b$

9 Simplify the following.

a $\left(\frac{\sqrt{2}}{3}\right)^2$ **b** $\left(\frac{4}{\sqrt{3}}\right)^2$

★**10** Simplify the following.

a $\sqrt{32} + \sqrt{8}$ **b** $\sqrt{32} \times \sqrt{8}$ **c** $\sqrt{27} \times \sqrt{18} \div \sqrt{3}$

HOMEWORK 8L

1 Show that

a $(3 + \sqrt{5})(2 + \sqrt{5}) = 11 + 5\sqrt{5}$ **b** $(3 - \sqrt{2})(3 + \sqrt{2}) = 7$

2 Expand and simplify where possible.

a $\sqrt{5}(3 - \sqrt{2})$ **b** $\sqrt{8}(3 - 4\sqrt{2})$ **c** $3\sqrt{8}(2\sqrt{2} + 4)$

d $(2 + \sqrt{3})(1 - \sqrt{3})$ **e** $(3 + \sqrt{5})(2 - \sqrt{5})$ **f** $(3 - \sqrt{2})(4 + 2\sqrt{2})$

3 Work out the missing lengths in these triangles, simplifying the answer where possible.

a

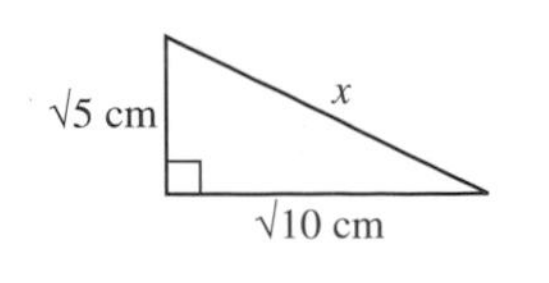

b

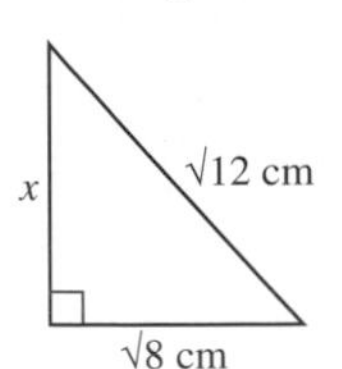

4 Calculate the area of these rectangles, simplifying your answer where possible.

a

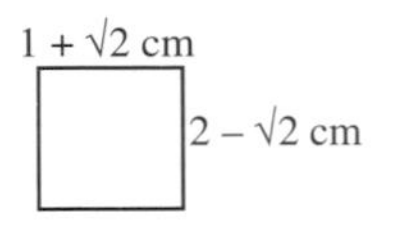

b

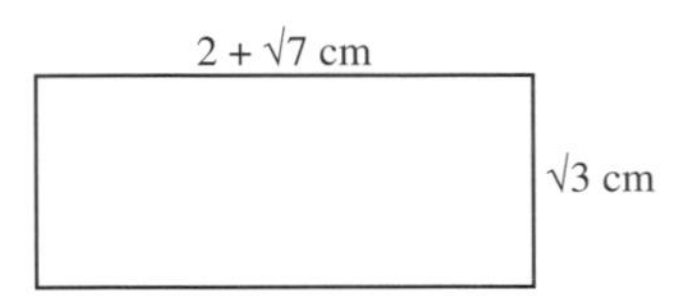

5 Rationalise the denominators of these expressions.

a $\frac{1}{\sqrt{7}}$ **b** $\frac{1}{\sqrt{8}}$ **c** $\frac{2}{\sqrt{5}}$ **d** $\frac{1}{2\sqrt{2}}$

e $\frac{5\sqrt{3}}{\sqrt{27}}$ **f** $\frac{\sqrt{8}}{\sqrt{3}}$ **g** $\frac{1+\sqrt{3}}{\sqrt{3}}$ **h** $\frac{3-\sqrt{2}}{\sqrt{8}}$

6 **a** Expand and simplify

i $(5+\sqrt{3})(5-\sqrt{3})$ **ii** $(6-\sqrt{2})(6+\sqrt{2})$

b Use your results from part **a** to rationalise the denominators of

i $\frac{5}{5-\sqrt{3}}$ **ii** $\frac{2+\sqrt{3}}{6+\sqrt{2}}$

★7 **a** If $m = 5$ and $n = -7$, write down the exact value of the following, leaving your answers as integers or fractions if you can, otherwise as an irrational number.

i $m^2 - n^2$ **ii** $m^{\frac{1}{2}}$ **iii** n^{-3}

b Simplify $(4-\sqrt{5})^2$ as far as possible, without using a calculator.

Chapter 9 Statistics 1

Homework 9A

1 Find **i** the mode, **ii** the median and **iii** the mean from each frequency table below.

a A survey of the collar sizes of all the male staff in a school gave these results.

Collar size	12	13	14	15	16	17	18
Number of staff	1	3	12	21	22	8	1

b A survey of the number of TVs in pupils homes gave these results.

Number of TVs	1	2	3	4	5	6	7
Frequency	12	17	30	71	96	74	25

2 A survey of the number of pets in each family of a school gave these results.

Number of pets	0	1	2	3	4	5
Frequency	28	114	108	16	15	8

a Each child at the school is shown in the data, how many children are at the school?
b Calculate the median number of pets in a family.
c How many families have less than the median number of pets?
d Calculate the mean number of pets in a family. Give your answer to 1 dp.

Homework 9B

1 The following table shows the daily sales of newspapers at a garage for a month.

Sun	Mon	Tue	Wed	Thur	Fri	Sat
125	87	63	91	42	115	169
112	74	43	68	32	154	147
148	98	72	71	56	122	153
110	129	87	75	47	143	194

Make a table showing the moving average using a seven-day span, and draw a graph to show the trend of newspaper sales over the month.

2 Geoff scored the following runs in the cricket season.

31 15 58 114 36 0 98 128 77 5
103 89 51 23 39 86 74 152 107 76

a Calculate the three innings moving average for the data and plot this on a graph.
b Comment on the scores through the season.

Homework 9C

1 Find for each table of values given below
i the modal group ii an estimate for the mean.

a

Score	0 – 20	21 – 40	41 – 60	61 – 80	81 – 100
Frequency	9	13	21	34	17

b

Cost (£)	0.00 – 10.00	10.01 – 20.00	20.01 – 30.00	30.01 – 40.00	40.01 – 60.00
Frequency	9	17	27	21	14

2 A survey was made to see how long casualty patients had to wait before seeing a doctor. The following table summarises the results for one shift.

Time (minutes)	0 – 10	11 – 20	21 – 30	31 – 40	41 – 50	51 – 60	61 – 70
Frequency	1	12	24	15	13	9	5

a How many patients were seen by a doctor in the survey of this shift?
b Estimate the mean waiting time taken per patient.
c Which average would the hospital use for the average waiting time?
d What percentage of patients did the doctors see within the hour?

HOMEWORK 9D

1 After a mental arithmetic test, all the results were collated for girls and boys as below.

Number correct	0 – 5	6 – 10	11 – 15	16 – 20	21 – 30
Boys	5	9	18	24	21
Girls	4	8	22	21	19

i Draw frequency polygons to illustrate the differences between the boys' scores and the girls' scores.

ii Estimate the mean score for boys and girls separately, and comment on the results.

2 In a survey of 500 families about the number of electronic games consoles they had in their home, the following results were obtained.

Number of consoles	0	1	2	3	4
Frequency	3	67	245	168	17

i What is the mean number of consoles per family?

ii Why is the data not particularly suitable for a frequency polygon or a histogram to be drawn?

HOMEWORK 9E

1 **a** The table shows the ages of 300 people at the cinema.

Age, x years	$0 \le x < 10$	$10 \le x < 15$	$15 \le x < 20$	$20 \le x < 30$	$30 \le x < 60$
Frequency	25	85	115	45	30

Draw a histogram to show the data.

b At another film show this was the distribution of ages

Age, x years	$15 \le x < 20$	$20 \le x < 30$	$30 \le x < 40$	$40 \le x < 60$	$60 \le x < 80$
Frequency	35	120	130	50	15

Draw a histogram to show this data.

c Comment on the differences between the distributions.

2 The table shows the times taken by 50 children to complete a multiplication square.

Time, s seconds	$10 \le s < 20$	$20 \le s < 25$	$25 \le s < 30$	$30 \le s < 40$	$40 \le s < 60$
Frequency	3	9	28	6	4

a Draw a histogram to show the data.

b Calculate an estimate of the mean of the data.

c Draw a vertical line on the histogram at the mean value.

d What is the significance of this line in relation to the size of the bars of the histogram?

3 For the histogram below

a write out the frequency diagram

b calculate an estimate of the mean of the distribution.

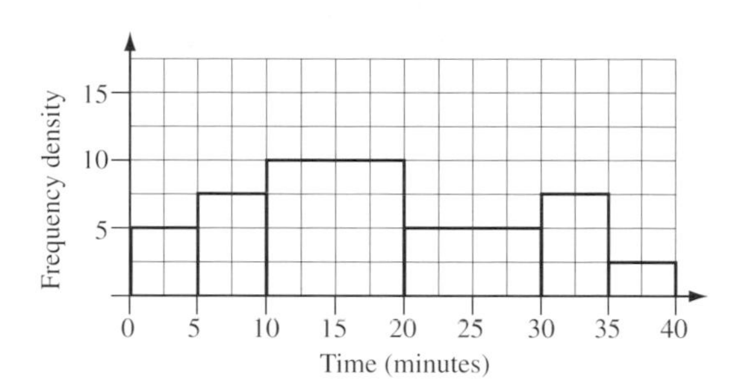

★4 The waiting times for customers at a supermarket checkout are shown in the table.

a Draw a histogram of these waiting times.

b Show an estimate of the median on your histogram. Show your working.

Waiting time (minutes)	Frequency
$0 \le x < 1$	15
$1 \le x < 3$	7
$3 \le x < 4$	12
$4 \le x < 5$	15
$5 \le x < 10$	12

Homework 9F

1 'People like the video hire centre to be open 24 hours a day.'

a To see whether this statement is true, design a data collection sheet which will allow you to capture data while standing outside a video hire centre.

b Does it matter at which time you collect your data?

2 The Youth Club wanted to know which types of activities it should plan, e.g. craft, swimming, squash, walking, disco etc.

a Design a data collection sheet which you could use to ask the pupils in your school which activities they would want in a Youth Club.

b Invent the first 30 entries on the chart.

★3 What types of film do your age group watch at the cinema the most? Is it comedy, romance, sci-fi, action, suspense or something else?

a Design a data collection sheet to be used in a survey of your age group.

b Invent the first thirty entries on your sheet.

Homework 9G

1 Design a questionnaire to test the following statement.
'Young people aged 16 and under will not tell their parents when they have been drinking alcohol, but the over 16s will always let their parents know.'

★2 'Boys will use the Internet almost everyday but girls will only use it about once a week.'
Design a questionnaire to test this statement.

★3 Design a questionnaire to test the following hypothesis.
'When you are in your twenties, you watch less TV than any other age group.'

4 While on holiday in Wales, I noticed that in the supermarkets there were a lot more women than men, and even then, the only men I did see were over 65.

a Write down a hypothesis from the above observation.

b Design a questionnaire to test your hypothesis.

Homework 9H

1 For a school project you have been asked to do a presentation of the timing of the school day. You decide to interview a sample of pupils. How will you choose those you wish to interview if you want your results to be reliable? Give three reasons for your decisions.

2 Comment on the reliability of the following ways of finding a sample.

a Asking the year 11 Religious Education option class about religion.

b Find out how many homes have microwaves by asking the first 100 pupils who walk through the school gates.

c Find the most popular Playstation game by asking a Year 7 form which their favourite is.

3 Comment on the way the following samples have been taken. For those that are not satisfactory, suggest a better way to find a more reliable sample.

a Bill wanted to find out what proportion of his school went to the cinema, so he obtained an alphabetical list of students and sent a questionnaire to every 10th person on the list.

b The council wants to know about sports facilities in an area so they sent a survey team to the local shopping centre one Monday morning.

c A political party wanted to know how much support they had in an area so they rang 500 people from the phone book in the evening.

4 Shameela made a survey of pupils in her school. The size of each year group in the school is shown below.

Year	Boys	Girls	Total
7	154	137	291
8	162	156	318
9	134	160	294
10	153	156	309
11	130	140	270
Total	**733**	**749**	**1482**

Claire took a sample of 150 pupils.

a Explain why this is a suitable size of sample.

b Draw up a table showing how many pupils of each sex and year she should ask if she wants to obtain a stratified sample.

★5 A train company attempted to estimate the number of people who travel by train in a certain town. They telephoned 200 people in the town one evening and asked, 'Have you travelled by train in the last week?' 32 people said 'Yes.' The train company concluded that 16% of the town's population travel by train. Give three criticisms of this method of estimation.

Chapter 10 Algebra 2

Homework 10A

Evaluate these expressions, writing them as simply as possible.

1 $3 \times 4t$ **2** $2 \times 5y$ **3** $4y \times 2$ **4** $3w \times 3$
5 $4t \times t$ **6** $6b \times b$ **7** $3w \times w$ **8** $6y \times 2y$
9 $5p \times p$ **10** $4t \times 32t$ **11** $5m \times 4m$ **12** $6t \times 4t$
13 $m \times 7t$ **14** $5y \times w$ **15** $8t \times q$ **16** $n \times 69t$
17 $5 \times 6q$ **18** $5f \times 2$ **19** $6 \times 3k$ **20** $5 \times 7r$
21 $t^2 \times t$ **22** $p \times p^2$ **23** $5m \times m^2$ **24** $3t^2 \times t$
25 $4n \times 2n^2$ **26** $5r^2 \times 4r$ **27** $t^2 \times t^2$ **28** $k^3 \times k^2$
29 $8n^2 \times 2n^3$ **30** $4t^3 \times 3t^4$ **31** $7a^4 \times 2a^3$ **32** $k^5 \times 3k^2$
33 $-k^2 \times -k$ **34** $-5y \times -2y$ **35** $-3d^2 \times -6d$ **36** $-2p^4 \times 6p^2$
37 $5mq \times q$ **38** $4my \times 3m$ **39** $4mt \times 3m$ **40** $5qp \times 2qp$

Homework 10B

Expand these expressions.

1 $3(4 + m)$ **2** $6(3 + p)$ **3** $4(4 - y)$ **4** $3(6 + 7k)$
5 $4(3 - 5f)$ **6** $2(4 - 23w)$ **7** $7(g + h)$ **8** $4(2k + 4m)$
9 $6(2d - n)$ **10** $t(t + 5)$ **11** $m(m + 4)$ **12** $k(k - 2)$
13 $g(4g + 1)$ **14** $y(3y - 21)$ **15** $p(7 - 8p)$ **16** $2m(m + 5)$
17 $3t(t - 2)$ **18** $3k(5 - k)$ **19** $2g(4g + 3)$ **20** $4h(2h - 3)$
21 $2t(6 - 5t)$ **22** $4d(3d + 5e)$ **23** $3y(4y + 5k)$ **24** $6m^2(3m - p)$
25 $y(y^2 + 7)$ **26** $h(h^3 + 9)$ **27** $k(k^2 - 4)$ **28** $3t(t^2 + 3)$
29 $5h(h^3 - 2)$ **30** $4g(g^3 - 3)$ **31** $5m(2m^2 + m)$ **32** $2d(4d^2 - d^3)$
33 $4w(3w^2 + t)$ **34** $3a(5a^2 - b)$ **35** $2p(7p^3 - 8m)$ **36** $m^2(3 + 5m)$
37 $t^3(t + 3t)$ **38** $g^2(4t - 3g^2)$ **39** $2t^2(7t + m)$ **40** $3h^2(4h + 5g)$

Homework 10C

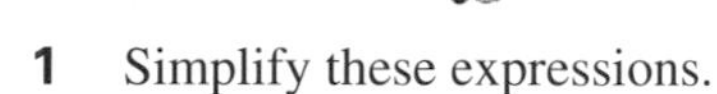

1 Simplify these expressions.

a $5t + 4t$ **b** $4m + 3m$ **c** $6y + y$ **d** $2d + 3d + 5d$
e $7e - 5e$ **f** $6g - 3g$ **g** $3p - p$ **h** $5t - t$
i $t^2 + 4t^2$ **j** $5y^2 - 2y^2$ **k** $4ab + 3ab$ **l** $5a^2d - 4a^2d$

2 Expand and simplify.

a $3(2 + t) + 4(3 + t)$ **b** $6(2 + 3k) + 2(5 + 3k)$ **c** $5(2 + 4m) + 3(1 + 4m)$
d $3(4 + y) + 5(1 + 2y)$ **e** $5(2 + 3f) + 3(6 - f)$ **f** $7(2 + 5g) + 2(3 - g)$

3 Expand and simplify.

a $4(3 + 2h) - 2(5 + 3h)$ **b** $5(3g + 4) - 3(2g + 5)$ **c** $3(4y + 5) - 2(3y + 2)$
d $3(5t + 2) - 2(4t + 5)$ **e** $5(5k + 2) - 2(4k - 3)$ **f** $4(4e + 3) - 2(5e - 4)$

4 Expand and simplify.

a $m(5 + p) + p(2 + m)$ **b** $k(4 + h) + h(5 + 2k)$ **c** $t(1 + 2n) + n(3 + 5t)$
d $p(5q + 1) + q(3p + 5)$ **e** $2h(3 + 4j) + 3j(h + 4)$ **f** $3y(4t + 5) + 2t(1 + 4y)$

5 Expand and simplify.

a $t(2t+5)+2t(4+t)$ **b** $3y(4+3y)+y(6y-5)$ **c** $5w(3w+2)+4w(3-w)$

d $4p(2p+3)-3p(2-3p)$ **e** $4m(m-1)+3m(4-m)$ **f** $5d(3-d)+d(2d-1)$

6 Expand and simplify.

a $5a(3b+2a)+a(2a^2+3c)$ **b** $4y(3w+y^2)+y(3y-4t)$

Homework 10D

Factorise the following expressions.

1 $9m+12t$	**2** $9t+6p$	**3** $4m+12k$	**4** $4r+6t$
5 $2mn+3m$	**6** $4g^2+3g$	**7** $4w-8t$	**8** $10p-6k$
9 $12h-10k$	**10** $4mp+2mk$	**11** $4bc+6bk$	**12** $8ab+4ac$
13 $3y^2+4y$	**14** $5t^2-3t$	**15** $3d^2-2d$	**16** $6m^2-3mp$
17 $3p^2+9pt$	**18** $8pt+12mp$	**19** $8ab-6bc$	**20** $4a^2-8ab$
21 $8mt-6pt$	**22** $20at^2+12at$	**23** $4b^2c-10bc$	**24** $4abc+6bed$

25 $6a^2+4a+10$ **26** $12ab+6bc+9bd$ **27** $6t^2+3t+at$

28 $96mt^2-3mt+69m^2t$ **29** $6ab^2+2ab-4a^2b$ **30** $5pt^2+15pt+5p^2t$

Factorise the following expressions where possible. List those which cannot factorise.

31 $5m-6t$	**32** $3m+2mp$	**33** t^2-5t	**34** $6pt+5ab$
35 $8m^2-6mp$	**36** a^2+c	**37** $3a^2-7ab$	**38** $4ab+5cd$
39 $7ab-4b^2c$	**40** $3p^2-4t^2$	**41** $6m^2t+9t^2m$	**42** $5mt+3pn$

Homework 10E

Expand the following expressions.

1 $(x+2)(x+5)$	**2** $(t+3)(t+2)$	**3** $(w+4)(w+1)$
4 $(m+6)(m+2)$	**5** $(k+2)(k+4)$	**6** $(a+3)(a+1)$
7 $(x+3)(x-1)$	**8** $(t+6)(t-4)$	**9** $(w+2)(w-3)$
10 $(f+1)(f-4)$	**11** $(g+2)(g-5)$	**12** $(y+5)(y-2)$
13 $(x-4)(x+3)$	**14** $(p-3)(p+2)$	**15** $(k-5)(k+1)$
16 $(y-3)(y+6)$	**17** $(a-2)(a+4)$	**18** $(t-4)(t+5)$
19 $(x-3)(x-2)$	**20** $(r-4)(r-1)$	**21** $(m-1)(m-7)$
22 $(g-5)(g-3)$	**23** $(h-6)(h-2)$	**24** $(n-2)(n-8)$
25 $(4+x)(3+x)$	**26** $(5+t)(4-t)$	**27** $(2-b)(6+b)$
28 $(7-y)(5-y)$	**29** $(3+p)(p-2)$	**30** $(3-k)(k-5)$

Homework 10F

Expand the following expressions.

1 $(3x+4)(4x+2)$	**2** $(2y+1)(3y+2)$	**3** $(4t+2)(3t+6)$
4 $(3t+2)(2t-1)$	**5** $(6m+1)(3m-2)$	**6** $(5k+3)(4k-3)$
7 $(4p-5)(3p+4)$	**8** $(6w+1)(3w+4)$	**9** $(3a-4)(5a+1)$
10 $(5r-2)(3r-1)$	**11** $(4g-1)(3g-2)$	**12** $(3d-2)(4d+1)$
13 $(3+4p)(5+4p)$	**14** $(3+2t)(5+3t)$	**15** $(2+5p)(3p+1)$
16 $(7+4t)(3-2t)$	**17** $(5+2n)(4-n)$	**18** $(3+4f)(5f-1)$
19 $(2-3q)(5+4q)$	**20** $(3-p)(2+3p)$	**21** $(5-3t)(4t+1)$

22 $(5-4r)(3-4r)$	**23** $(4-x)(1-5x)$	**24** $(2-7m)(2m-3)$
25 $(x+y)(3x+5y)$	**26** $(4y+t)(3y-4t)$	**27** $(5x-3y)(5x+y)$
28 $(x-2y)(x-3y)$	**29** $(4m-3p)(m+5p)$	**30** $(t-4k)(3t-k)$

HOMEWORK 10G

Try to spot the pattern in each of the following expressions so that you can immediately write down the expansion.

1 $(x+1)(x-1)$	**2** $(t+2)(t-2)$	**3** $(y+3)(y-3)$
4 $(2m+3)(2m-3)$	**5** $(4k-3)(4k+3)$	**6** $(5h-1)(5h+1)$
7 $(3+2x)(3-2x)$	**8** $(7+2t)(7-2t)$	**9** $(4-5y)(4+5y)$
10 $(a+b)(a-b)$	**11** $(3t+k)(3t-k)$	**12** $(m-3p)(m+3p)$
13 $(8k+g)(8k-g)$	**14** $(ac+bd)(ac-bd)$	**15** $(x^2+y^2)(x^2-y^2)$

HOMEWORK 10H

Expand the following squares.

1 $(x+4)^2$	**2** $(m+3)^2$	**3** $(5+t)^2$	**4** $(2+p)^2$
5 $(m-2)^2$	**6** $(t-4)^2$	**7** $(3-m)^2$	**8** $(6-k)^2$
9 $(2x+1)^2$	**10** $(3t+2)^2$	**11** $(1+4y)^2$	**12** $(2+m)^2$
13 $(3t-2)^2$	**14** $(2x-1)^2$	**15** $(1-4t)^2$	**16** $(5-4r)^2$
17 $(a+b)^2$	**18** $(x-y)^2$	**19** $(3t+y)^2$	**20** $(m-2n)^2$
21 $(x+3)^2-4$	**22** $(x-4)^2-25$	**23** $(x+5)^2-36$	**24** $(x-1)^2-1$

HOMEWORK 10I

Factorise the following.

1 x^2+7x+6	**2** t^2+4t+4	**3** $m^2+11m+10$	**4** $k^2+11k+24$
5 $p^2+10p+24$	**6** $r^2+11r+18$	**7** $w^2+9w+18$	**8** $x^2+8x+12$
9 $a^2+13a+12$	**10** $k^2-10k+21$	**11** $f^2-22f+21$	**12** $b^2+35b+96$
13 t^2+5t+6	**14** m^2-5m+4	**15** $p^2-7p+10$	**16** $x^2-13x+36$
17 $c^2-12c+32$	**18** $t^2-15t+36$	**19** $y^2-14y+48$	**20** $j^2-19j+48$
21 $p^2+8p+15$	**22** y^2+y-6	**23** t^2+7t-8	**24** $x^2+9x-10$
25 m^2-m-12	**26** r^2+6r-7	**27** $n^2-7n-18$	**28** $m^2-20m-44$
29 $w^2-5w-24$	**30** t^2+t-90	**31** x^2-x-72	**32** $t^2-18t-63$
33 d^2-2d+1	**34** $y^2+29y+100$	**35** $t^2-10t+16$	**36** $m^2-30m+81$
37 $x^2-30x+144$	**38** $d^2-4d-12$	**39** t^2+t-20	**40** q^2+q-56
41 p^2-p-2	**42** $v^2-2v-35$	**43** t^2-4t+3	**44** m^2+3m-4

HOMEWORK 10J

Each of these is the difference of two squares. Factorise them.

1 x^2-81	**2** t^2-36	**3** $4-x^2$	**4** $81-t^2$
5 k^2-400	**6** $64-y^2$	**7** x^2-y^2	**8** a^2-9b^2
9 $9x^2-25y^2$	**10** $9x^2-16$	**11** $100t^2-4w^2$	**12** $36a^2-49b^2$

★**13** Simplify $\dfrac{2a-3}{4a^2-9}$

Homework 10K

Solve these equations.

1 $(x + 3)(x + 2) = 0$ **2** $(t + 4)(t + 1) = 0$ **3** $(a + 5)(a + 3) = 0$
4 $(x + 4)(x - 1) = 0$ **5** $(x + 2)(x - 5) = 0$ **6** $(t + 3)(t - 4) = 0$
7 $(x - 2)(x + 1) = 0$ **8** $(x - 1)(x + 4) = 0$ **9** $(a - 6)(a + 5) = 0$
10 $(x - 2)(x - 5) = 0$ **11** $(x - 2)(x - 1) = 0$ **12** $(a - 2)(a - 6) = 0$

First factorise, then solve these equations.

13 $x^2 + 6x + 5 = 0$ **14** $x^2 + 9x + 18 = 0$ **15** $x^2 - 7x + 8 = 0$
16 $x^2 - 4x + 21 = 0$ **17** $x^2 + 3x - 10 = 0$ **18** $x^2 + 2x - 15 = 0$
19 $t^2 - 4t - 12 = 0$ **20** $t^2 - 3t - 18 = 0$ **21** $x^2 + x - 2 = 0$
22 $x^2 - 4x + 4 = 0$ **23** $m^2 - 10m + 25 = 0$ **24** $t^2 - 10t + 16 = 0$
25 $t^2 + 7t + 12 = 0$ **26** $k^2 - 3k - 18 = 0$ **27** $a^2 - 20a + 64 = 0$

Homework 10L

Factorise the following expressions.

1 $3x^2 + 4x + 1$ **2** $3x^2 - 2x - 1$ **3** $4x^2 + 8x + 3$ **4** $2x^2 + 7x + 3$
5 $15x^2 + 13x + 2$ **6** $4x^2 + 4x - 3$ **7** $6x^2 - 7x + 2$ **8** $8x^2 - 8x - 6$
9 $8x^2 - 13x - 6$ **10** $6x^2 - 13x + 2$

★**11** Factorise **a** $6x^2 - 2x$ **b** $6x^2 + 11x - 2$

Homework 10M

Give your answers either in rational form or as mixed numbers.

1 Solve the following equations.
a $2x^2 + 5x + 2 = 0$ **b** $7x^2 + 8x + 1 = 0$ **c** $4x^2 + 3x - 7 = 0$
d $6x^2 + 13x + 5 = 0$ **e** $6x^2 + 7x + 2 = 0$

2 Solve the following equations.
a $x^2 - x = 6$ **b** $2x(4x + 7) = -3$ **c** $(x + 3)(x - 4) = 18$
d $11x = 21 - 2x^2$ **e** $(2x + 3)(2x - 3) = 9x$

★**3** **a** Simplify $\dfrac{x^2 - 9}{3x - 9}$ **b** Solve the equation $12x^2 - 25x + 12 = 0$

Homework 10N

Solve the following equations using the quadratic formula. Give your answers to 2 dp.

1 $3x^2 + x - 5 = 0$ **2** $2x^2 + 4x + 1 = 0$ **3** $x^2 - x - 7 = 0$
4 $3x^2 + x - 1 = 0$ **5** $3x^2 + 7x + 3 = 0$ **6** $2x^2 + 11x + 1 = 0$
7 $2x^2 + 5x + 1 = 0$ **8** $x^2 + 2x - 9 = 0$ **9** $x^2 + 2x - 6 = 0$

★**10** Solve the equation $x^2 = 5x + 7$, giving your answers correct to 3 significant figures.

HOMEWORK 10P

1 Write an equivalent expression in the form $(x \pm a)^2 - b$.
  a $x^2 + 10x$ **b** $x^2 + 18x$ **c** $x^2 - 8x$ **d** $x^2 + 20x$ **e** $x^2 + 7x$

2 Write an equivalent expression in the form $(x \pm a)^2 - b$.
  a $x^2 + 10x - 1$ **b** $x^2 + 18x - 5$ **c** $x^2 - 8x + 3$ **d** $x^2 - 5x - 1$

3 Solve the following equations by completing the square. Leave your answers in surd form where appropriate. The answers to Question **2** will help.
  a $x^2 + 10x - 1 = 0$ **b** $x^2 + 18x - 5 = 0$ **c** $x^2 - 8x + 3 = 0$
  d $x^2 + 20x + 7 = 0$ **e** $x^2 - 5x - 1 = 0$

4 Solve $x^2 + 8x - 3 = 0$ by completing the square. Give your answers to 2 dp.

★5 **a** Write the equation $x^2 + 4x - 6$ in the form $(x + a)^2 - b$.
  b Hence or otherwise, solve the equation $x^2 + 4x - 6 = 0$, leaving your answer in surd form.

HOMEWORK 10Q

Solve the following equations. Where there is a solution, give your answer correct to 2 dp.
  1 $5x^2 + 2x = 1$ **2** $3x^2 - 4 = 5x$ **3** $3x^2 = 5x - 7$
  4 $5x^2 = 7 - 2x$ **5** $(2 - x) = (2x - 1)^2$ **6** $(x - 2)^2 = 11$

★7 Solve the equation $3x + 7 = 2x^2$, giving your answers to 2 dp.

HOMEWORK 10R

1 Solve the following equations. Give your answers in the form $a \pm \sqrt{b}$.
  a $x^2 - 6x - 4 = 0$ **b** $x^2 + 2x - 10 = 0$ **c** $x^2 + 6x - 8 = 0$
  d $x^2 + 4x - 6 = 0$ **e** $x^2 - 2x - 2 = 0$

2 Solve the following equations. Give your answers in surd form.
  a $2x^2 + 4x - 3 = 0$ **b** $3x^2 + 4x - 2 = 0$ **c** $2x^2 + 5x - 6 = 0$
  d $3x^2 - 5x - 6 = 0$ **e** $3x^2 + x - 5 = 0$

★3 $x + \frac{1}{x} = 3$
  a Show that this equation can be rearranged as $x^2 - 3x + 1 = 0$.
  b Solve this equation to find the values of x in surd form.

HOMEWORK 10S

1 The sides of a right-angled triangle are $3x$, $(x + 1)$ and $(4x - 3)$. Find the actual dimensions of the triangle.

2 The length of a rectangle is 3 m more than its width. Its area is 130 m². Find the actual dimensions of the rectangle.

3 Solve the equation $x + \dfrac{2}{x} = 5$. Give your answers correct to 2 decimal places.

4 Solve the equation $3x + \dfrac{2}{x} = 7$.

5 The area of a triangle is $24\,\text{cm}^2$. The base is 8 cm longer than the height. Use this information to set up a quadratic equation. Solve the equation to find the length of the base.

6 On a journey of 210 km, the driver of a train calculates that if he were to increase his average speed by 10 km/h, he would take 30 minutes less. Find his average speed.

7 After a price increase for bananas of 25p per kilogram I can buy 2 kilograms less for £6 than I could last week. How much do bananas cost this week?

★8 Gareth took part in a 26-mile road race.

a He ran the first 15 miles at an average speed of x mph. He ran the last 11 miles at an average speed of $(x - 2)$ mph. Write down an expression, in terms of x, for the time he took to complete the 26-mile race.

b Gareth took four hours to complete the race. Using your answer to part **a**, form an equation in terms of x.

c **i** Simplify your equation and show that it can be written as $2x^2 - 17x + 15 = 0$

ii Solve the equation and obtain Gareth's average speed over the first 15 miles.

Homework 10T

1 Simplify each of these.

a $\dfrac{x}{2} + \dfrac{x}{5}$ **b** $\dfrac{3x}{5} + \dfrac{x}{4}$ **c** $\dfrac{x-2}{3} + \dfrac{x+3}{2}$ **d** $\dfrac{x-4}{3} + \dfrac{2x-3}{5}$

2 Simplify each of these.

a $\dfrac{x}{2} - \dfrac{x}{5}$ **b** $\dfrac{3x}{5} - \dfrac{x}{4}$ **c** $\dfrac{x-2}{3} - \dfrac{x-3}{2}$ **d** $\dfrac{x-4}{5} - \dfrac{2x-3}{5}$

3 Simplify.

a $\dfrac{x}{2} \times \dfrac{x}{5}$ **b** $\dfrac{2x}{9} \times \dfrac{3x}{8}$ **c** $\dfrac{x-2}{3} \times \dfrac{9}{x-3}$ **d** $\dfrac{x-4}{12} \times \dfrac{6}{x^2-4x}$

4 Simplify.

a $\dfrac{x}{2} \div \dfrac{x}{5}$ **b** $\dfrac{2x}{9} \div \dfrac{3y}{15}$ **c** $\dfrac{x-5}{9} \div \dfrac{3}{x-1}$ **d** $\dfrac{x-4}{12} \div \dfrac{x^2-4x}{6}$

5 Solve the following equations.

a $\dfrac{x+1}{3} + \dfrac{x+3}{4} = 4$ **b** $\dfrac{x+2}{4} + \dfrac{x+1}{7} = 1$ **c** $\dfrac{4x+1}{2} - \dfrac{x+2}{4} = 7$

6 Show that the algebraic fractions simplify to the given expression.

a $\dfrac{3}{x+1} + \dfrac{4}{x+3} = 2$ simplifies to $2x^2 + x - 7 = 0$

b $\dfrac{5}{x+2} - \dfrac{3}{x+1} = 1$ simplifies to $x^2 + x + 3 = 0$

c $\dfrac{2}{3x-1} - \dfrac{3}{x-2} = 2$ simplifies to $6x^2 - 7x + 5 = 0$

7 Solve the following equations.

a $\frac{3}{x+1} + \frac{4}{x+2} = 2$ b $\frac{3}{x-2} + \frac{7}{x+2} = 2$ c $\frac{16}{3x-1} - \frac{4}{x+1} = 1$

★8 a Simplify the expression $\frac{2x}{5} + \frac{x}{4}$

b Hence or otherwise, solve the equation $\frac{2x}{5} + \frac{x}{4} = 18$

HOMEWORK 10U

1 Solve the linear simultaneous equations using the 'substitution' method.

a $x - y = 2$
$3x + y = 14$

b $2x - 5y = 1$
$5x + 3y = 18$

2 Solve the pairs of simultaneous equations.

a $xy = 12$
$2x = y + 10$

b $x^2 + y^2 = 40$
$y = x + 4$

c $y^2 - 1 = x^2 + xy$
$y = 4 - 2x$

d $x^2 - xy + y^2 = 13$
$y = 2x - 2$

★3 The circle $x^2 + y^2 = 20$ and the line $y = 2x$ are shown on the graph. Use an algebraic method to find the point P where the line and the circle intersect.

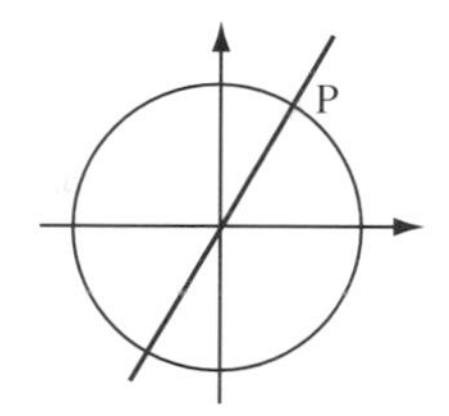

Chapter 11 Real-life graphs

HOMEWORK 11A

1 Joe was travelling in his car to meet his girlfriend. He set off from home at 9.00 pm, and stopped on the way for a break. This distance–time graph illustrates his journey.

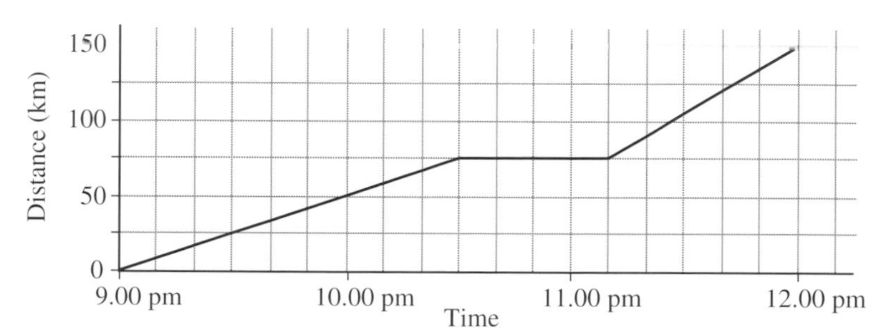

a At what time did he

i stop for his break ii set off after his break iii get to his meeting place?

b At what average speed was he travelling

i over the first hour ii over the last hour iii for the whole of his journey?

2 A taxi set off from Hellaby to pick up Jean. It then went on to pick up Jeans's parents. It then travelled further, dropping them all off at a shopping centre. The taxi then went on a further 10 km to pick up another party and took them back to Hellaby. This distance–time graph illustrates the journey.

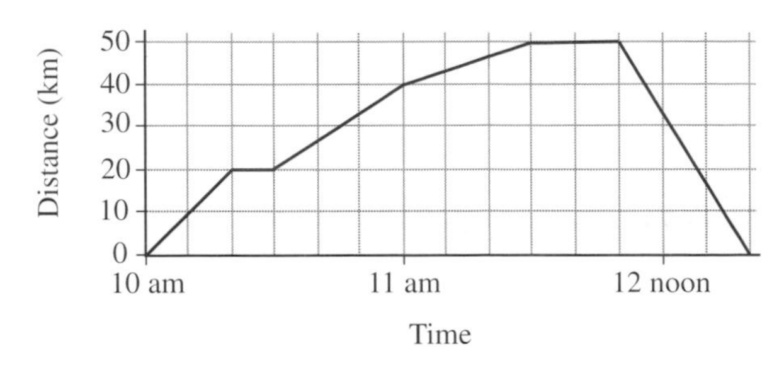

a How far from Hellaby did Jean's parents live?

b How far from Hellaby is the shopping centre?

c What was the average speed of the taxi while only Jean was in the taxi?

d What was the average speed of the taxi back to Hellaby?

3 Grandad took his grandchildren out for a trip. He set off at 1.00 pm and travelled, for half an hour, away from Norwich at the average speed of 60 km/h. They stopped to look at the sea and have an ice cream. At two o'clock, they set off again, travelling for a quarter of an hour at an average speed of 80 km/h. Here they stopped to play on the sand for half an hour. Grandad then drove the grandchildren back home, at an average speed of 50 km/h. Draw a travel graph to illustrate this story. Use a horizontal axis to represent time from 1 pm to 4 pm, and a vertical scale from 0 km to 50 km.

Homework 11B

1 Calculate the gradient of each line.

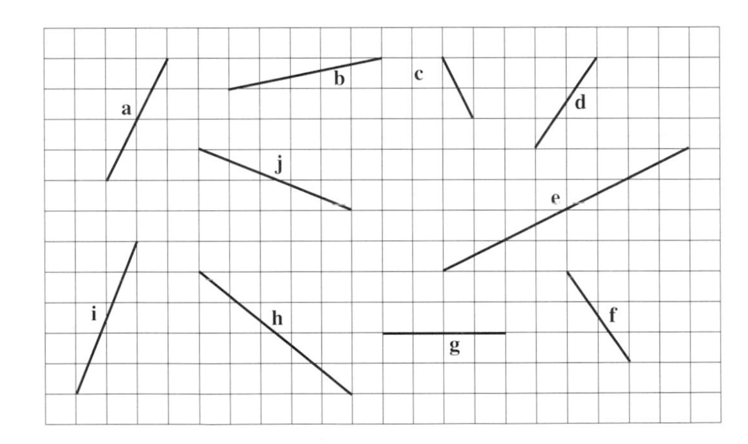

2 Calculate the average speed of the journey represented by each line in the following diagrams. The gradient of each line is the speed.

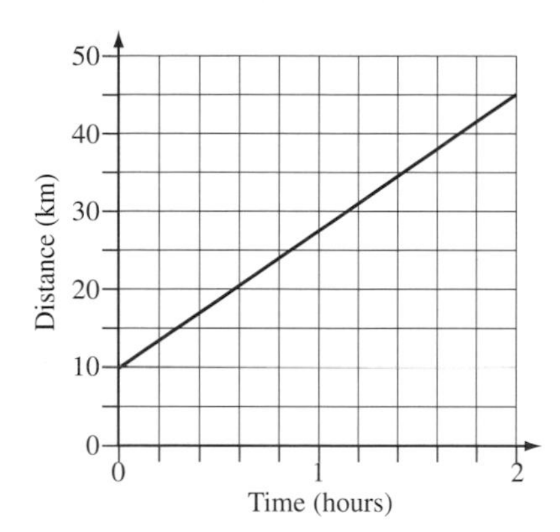

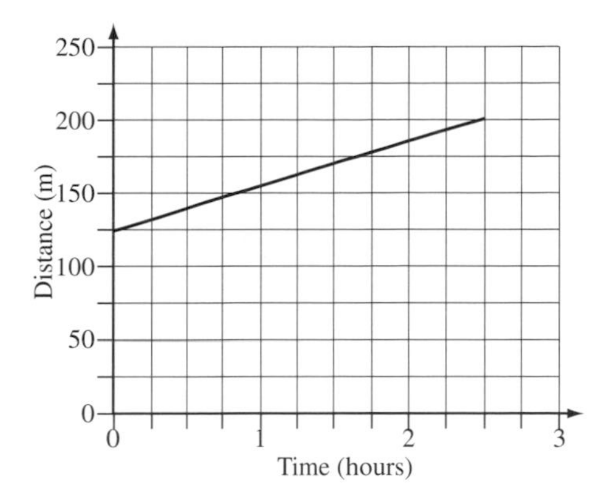

3 This is a conversion graph between ounces and grams.

a Calculate the gradient of the line.

b Use the graph to find the number of grams equivalent to 1 ounce.

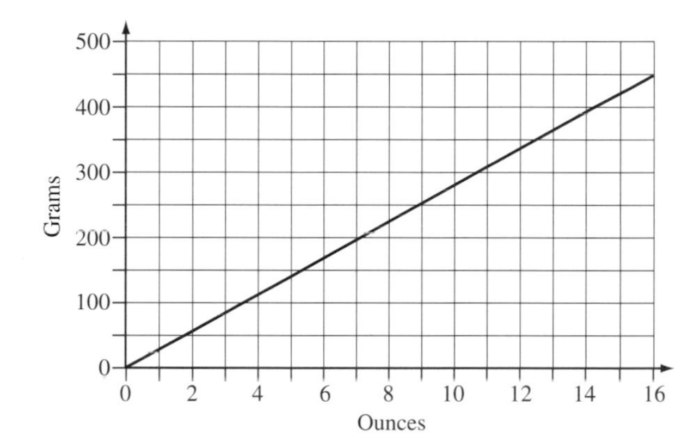

HOMEWORK 11C

1 The diagram shows the velocity of a car over 10 seconds.
Calculate the acceleration

a over the first 2 seconds

b after 6 seconds.

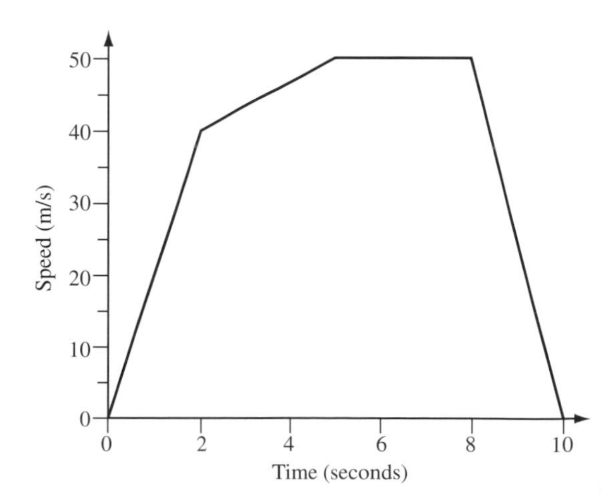

2

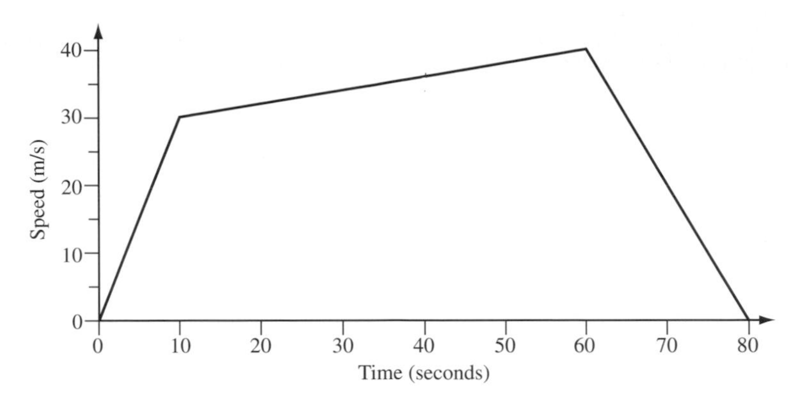

The diagram shows the velocity–time graph for a short train journey between stops. Find

a the acceleration over the first 10 seconds

b the deceleration over the last 20 seconds.

3 Starting from rest (zero velocity), a car travels as indicated below.

- Accelerates at a constant rate over 10 seconds to reach 20 m/s.
- Keeps this velocity for 30 seconds.
- Accelerates over the next 10 seconds to reach 30 m/s.
- Steadily slows down to reach rest (zero velocity) over the next 20 seconds.

a Draw the velocity–time graph.

b Calculate the deceleration over the last 20 seconds.

★4 The graph below represents the journey of a train.

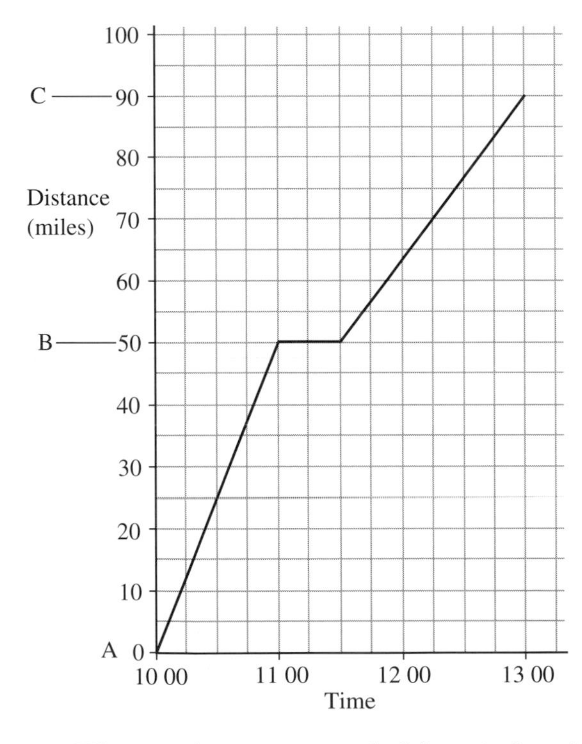

a What was the average speed of the train from A to B?

b How long did the train wait at B?

c Another train starts from C at 11.00 and travels non-stop to A at an average speed of 60 mph. Draw the graph of its journey on a copy of the graph.

d Write down how far from A the trains were when they passed each other.

HOMEWORK 11D

1 The following is a simplified model of how a savings scheme would work.
£100 is invested at the start of a year. At the end of the first year 7% interest is added to this and another £100 is invested. This gives a total of £207 at the start of the second year. At the end of the second year 7% is added and another £100 is invested. This continues for another 10 years.

a Calculate the amount of money in the scheme at the start of each year.

b Draw a graph of money in scheme against year. (First point will be (0, 100), second point will be (2, 207). Take the horizontal axis as 'Years' from 0 to 10. Take the vertical axis as 'Savings' from £0 to £1500.)

2 Flexible loans are loans where the amount paid back can vary. People usually take these out when they cannot pay a lot back at first but know that their income will increase so that they can pay a lot more back as time goes on. The Jones take out a £1000 loan at 8% interest. They pay back £50 after the first year, £75 after the second year, £100 after the third year and so on.
This is the calculation of what they owe for the first 2 years.

First year interest £1000 × 1.08 = £1080.
Amount owed at end of first year £1080 – £50 = £1030 owed.
Second year interest £1030 × 1.08 = £1112.40.
Amount owed at end of second year £1112.40 – 75 = £1037.40.

a Continue the calculation until the debt is paid off.

b Draw a graph of time (years) against amount owed.

3 Draw a graph of the depth of water in each of these containers as it is filled steadily.

a 

b 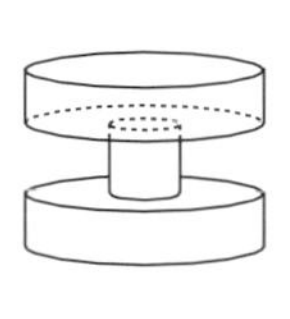

c 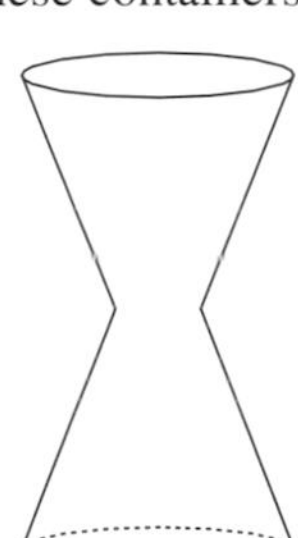

d 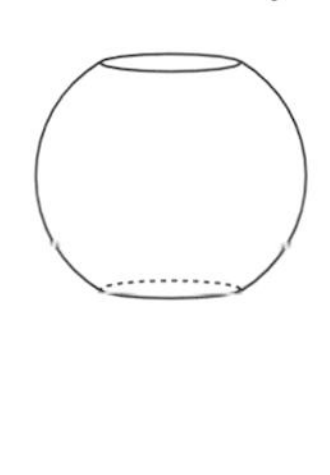

★4 a Each of the four graphs below represents **one** of the following **five** situations.

Situation A The first 4 seconds of a sprinter's 100 metre race.

Situation B A train travelling at a constant 15 m/s taking 4 seconds to pass through a tunnel.

Situation C A ball thrown vertically upwards and caught by the thrower 4 seconds later.

Situation D A ball rolling from rest down a smooth slope for 4 seconds.

Situation E A lift setting off from one floor and stopping 4 seconds later at the next floor.

1

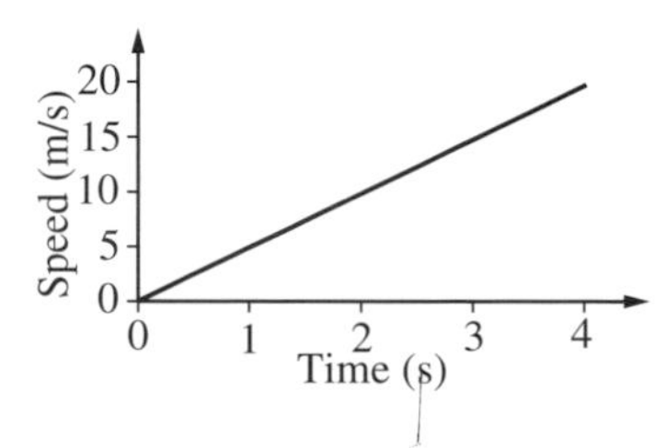

2

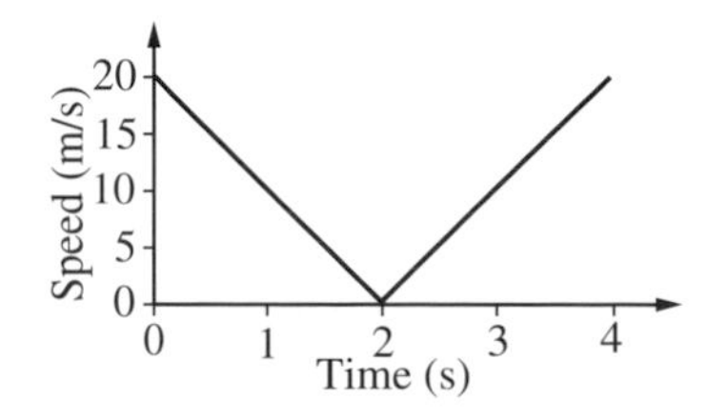

3

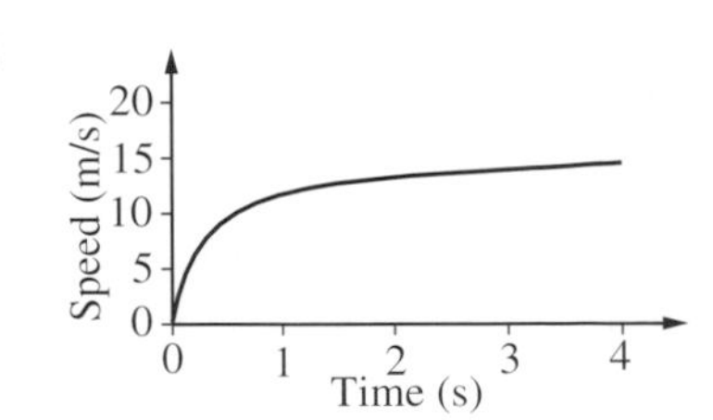

4

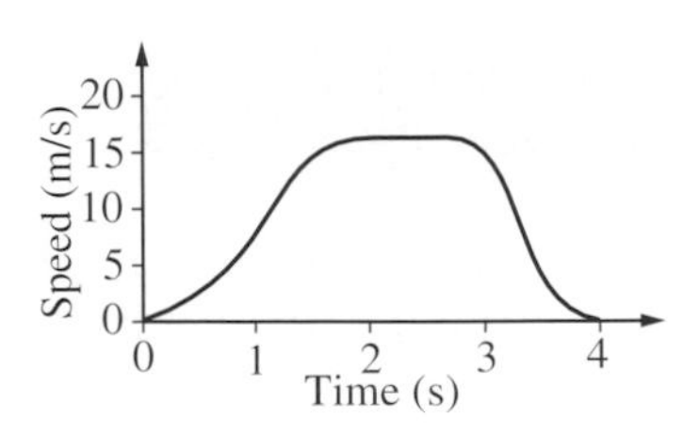

Match each graph with one of the five situations.

b One of the situations A, B, C, D or E does not match any of the four graphs drawn in part **a**. Sketch the speed–time graph for this situation.

Chapter 12 Similarity

Homework 12A

1 These diagrams are drawn to scale. What is the scale factor of the enlargement in each case?

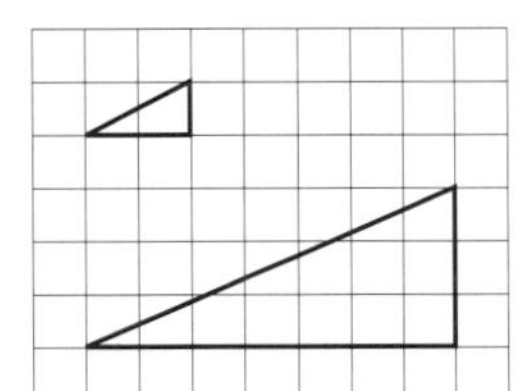

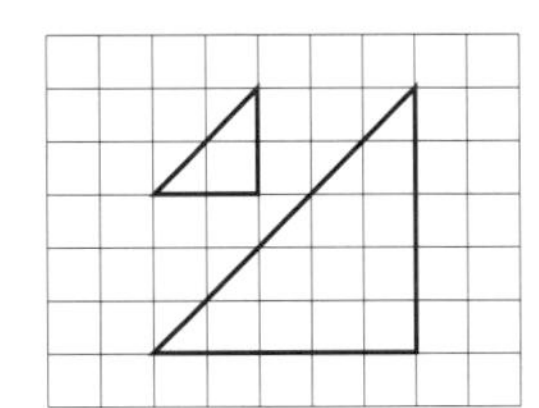

2 **a** Explain why these two shapes are similar.

b Give the ratio of the sides.

c Which angle corresponds to angle C?

d Which side corresponds to side QP?

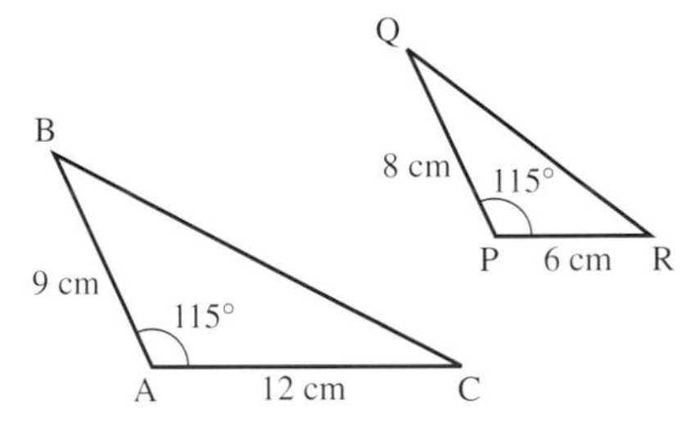

3 In the diagrams below, find the lengths of the sides marked x. Each pair of shapes is similar but not drawn to scale.

a

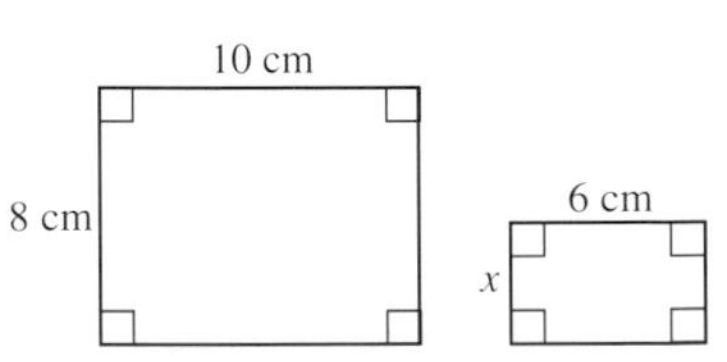

b

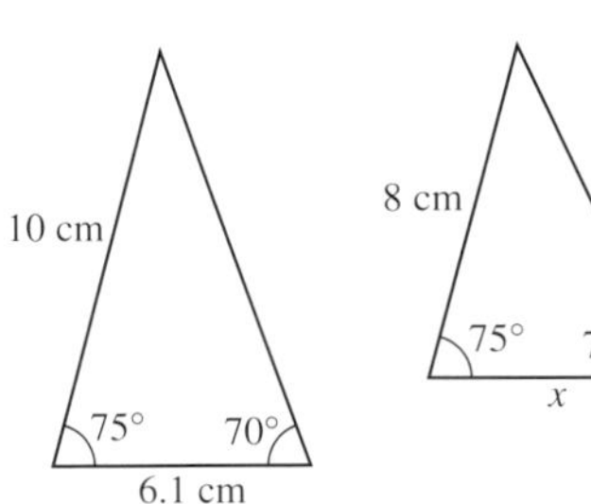

HOMEWORK 12B

1 In each of the cases below, state a pair of similar triangles and find the length marked x. Separate the similar triangles if it makes it easier for you.

a

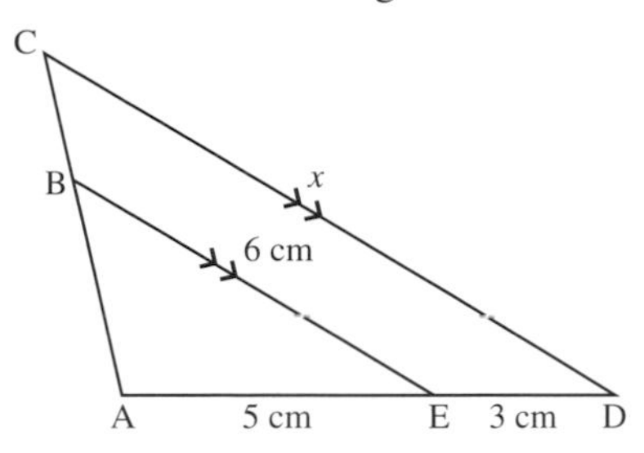

b

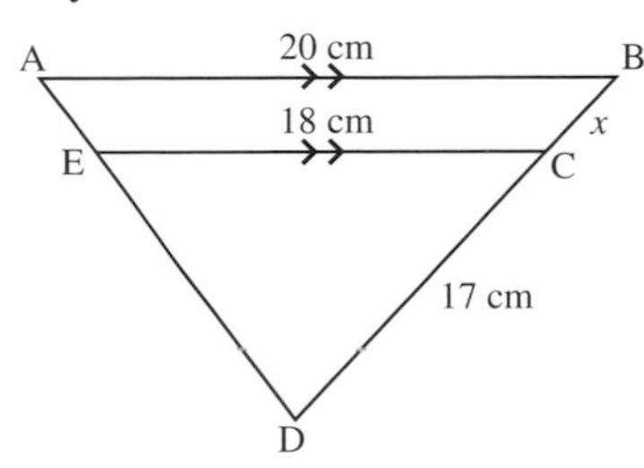

2 In the diagrams below, find the lengths of the sides marked x and y.

a

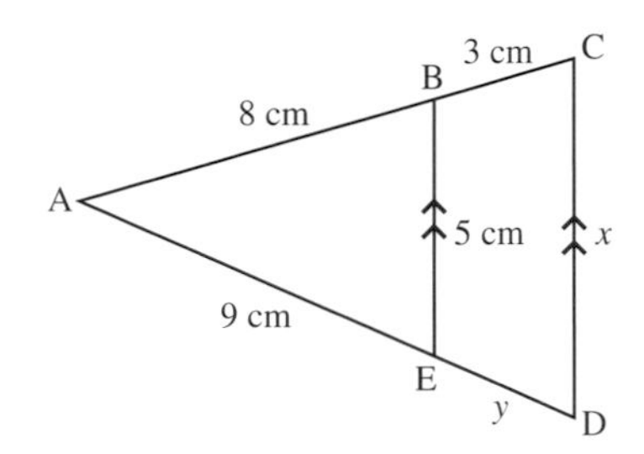

b

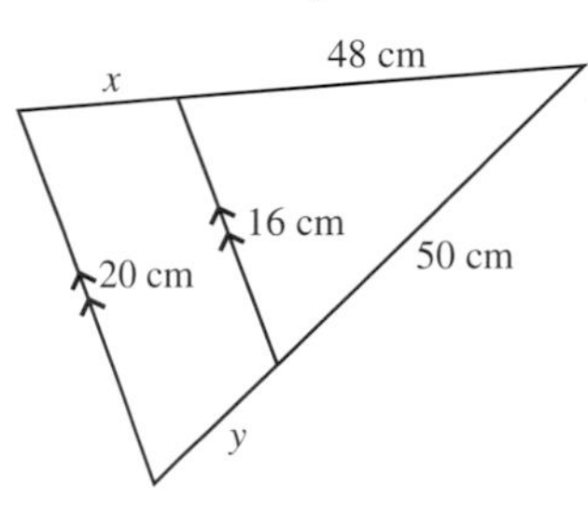

3 Find the height of a lamppost which casts a shadow of 2.1 metres when at the same time a man of height 158 cm casts a shadow of 90 cm.

4 Amy is 115 cm tall, she notices that when her mum, who is 175 cm tall, stands $2\frac{1}{2}$ metres away from her she looks as tall as the Cutlers Hall. Amy is standing 30 metres away from the Cutlers Hall when she notices this. How high is the Cutlers Hall?

HOMEWORK 12C

Find the lengths x and y in the diagrams below.

1

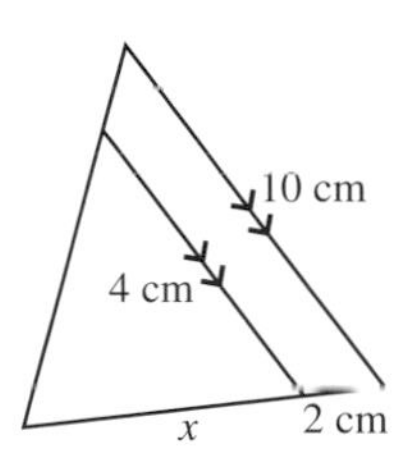

2

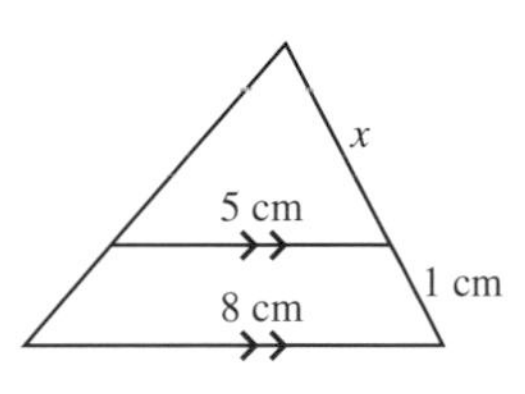

3

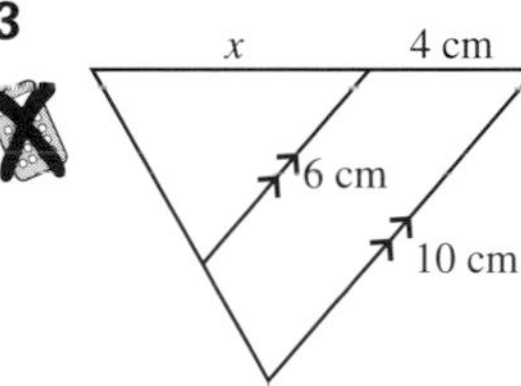

4

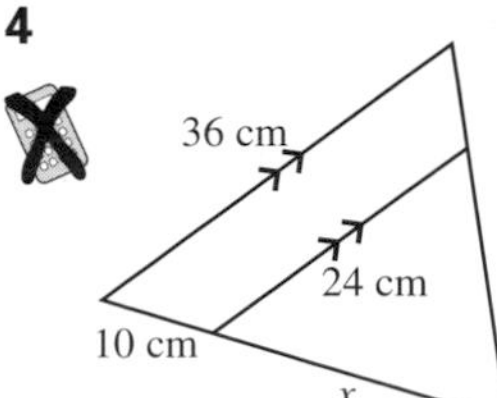

5

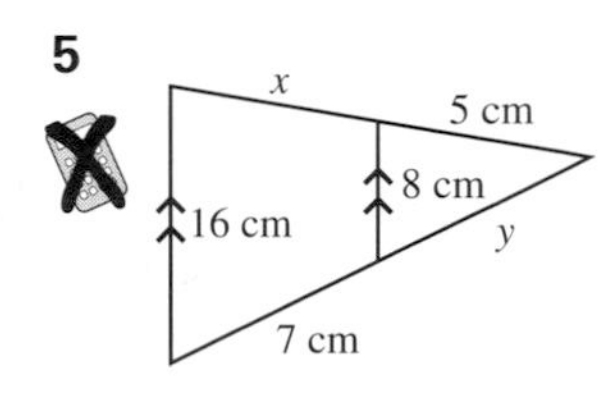

6

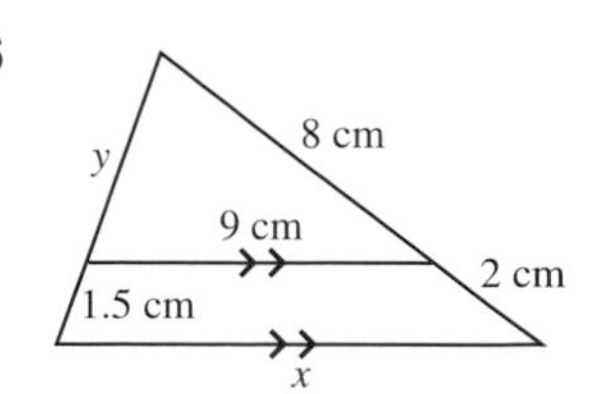

7

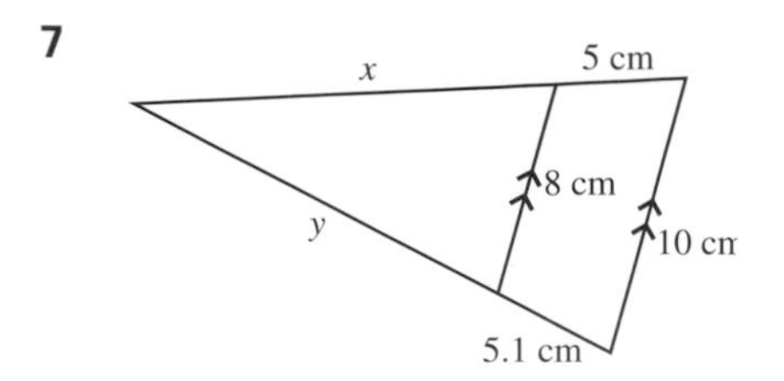

8

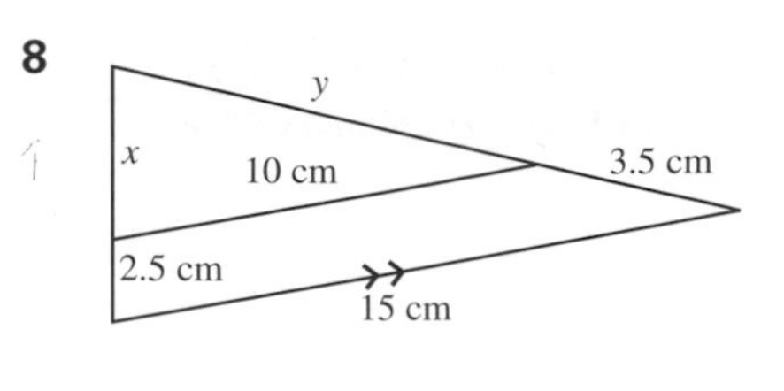

HOMEWORK 12D

1 The length ratio between two similar solids is 3 : 7.

a What is the area ratio between the solids?

b What is the volume ratio between the solids?

2 Copy and complete this table.

Linear scale factor	Linear ratio	Linear fraction	Area scale factor	Volume scale factor
4	1 : 4	$\frac{4}{1}$		
$\frac{1}{2}$				
	10 : 1			
			36	
				125

3 A shape has an area of 20 cm². What is the area of a similar shape whose lengths are four times the corresponding lengths of the first shape.

4 A brick has a volume of 400 cm³. What would be the volume of a similar brick whose lengths are

a three times the corresponding lengths of the first brick

b five times the corresponding lengths of the first brick?

5 A can of paint, 12 cm high, holds five litres of paint. How much paint would go into a similar can which is 30 cm high?

6 A model statue is 15 cm high and has a volume of 450 cm³. The real statue is 4.5 m high. What is the volume of the real statue? Give your answer in m³.

7 A triangle has sides of 5, 12 and 13 cm. Its area is 30 cm². How long are the sides of a similar triangle that has an area of 270 cm²?

8 A cuboid with one side of r cm has a volume of x cm³. What is the volume of a similar cuboid with one side of $3r$ cm?

9 What is **a** the linear scale factor and **b** the volume scale factor that corresponds to an area scale factor of $4a^2$

★10 A model motorbike is similar to the real bike in every detail. The scale is 1 : 10.

a The real bike's wheels have a diameter of 50 cm. What is the diameter of the model bike's wheels.

b The petrol tank on the real bike holds 12 litres. How much will the petrol tank on the model bike hold?

c The seat on the model bike took 20 cm² of leather to cover it. How much leather will be needed to cover the seat on the real bike?

HOMEWORK 12E

1 A firm produces three sizes of paper. Their areas are $600\,cm^2$, $900\,cm^2$ and $1200\,cm^2$. The dimensions of the smaller size are 20 cm by 30 cm. Calculate the dimensions of the other two sizes of paper.

2 A firm makes similar bottles in three different sizes: small, medium and large. The volumes are

small $330\,cm^3$ medium $1000\,cm^3$ large $2000\,cm^3$

a The medium bottle is 20 cm high. Find the heights of the other two bottles.

b The labels on the bottles are also similar. The large bottle has a label with an area of $100\,cm^2$. Find the areas of the labels on the other two bottles.

3 It takes 6 litres of paint to cover a wall with an area of 12 square metres. How large an area would 24 litres of paint cover?

4 It takes 30 minutes for 3 towels to dry on a washing line. How long would it take 6 similar towels to dry ?

5 It takes 1 kg of grass seed to cover a lawn that is 20 metres long. How much seed will be needed to cover a similar shaped lawn that is 10 metres long.

6 A model yacht has a mast 40 cm high. The real yacht has a mast 4 metres high.

a The sail on the model yacht has an area of $600\,cm^2$. What is the area of the real sail. Answer in m^2.

b The real yacht has a hull volume of $20\,m^3$. What is the hull volume of the model yacht? Answer in cm^3.

7 A triangle is enlarged as shown from the origin as centre.
The point A(2, 1) is transformed to A′(6, 3).
The area of the original triangle is $4\,cm^2$.
What is the area of the transformed triangle?

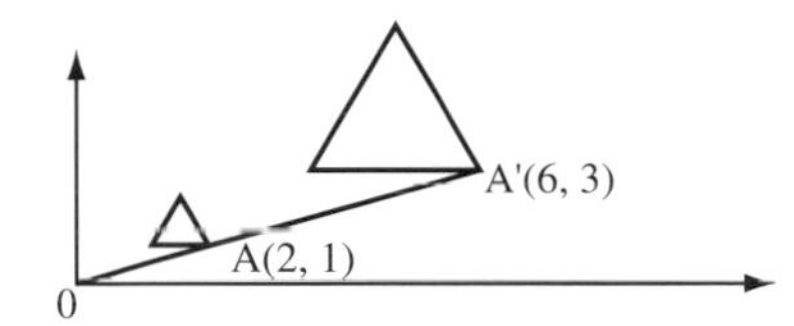

★8 The ratio of the height of P to the height of Q is 5 : 4. The volume of P is $150\,cm^3$.
Calculate the volume of Q.

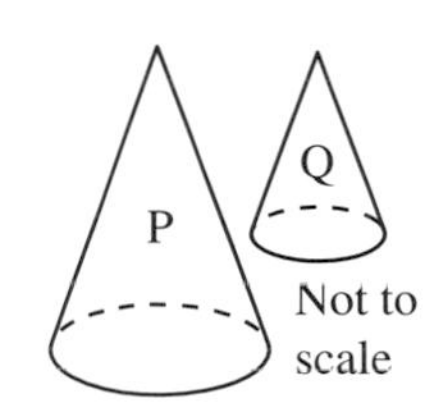

Chapter 13 Trigonometry

Homework 13A

1 A child's toy consists of a ball that fits into a cone.
The ball has a radius of 3 cm. The base angle of the cone is 38°.
Find

a AB **b** OB **c** AC

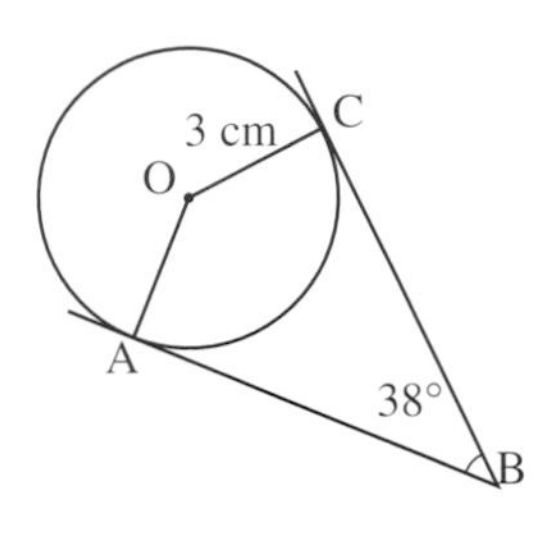

2 From the top of a building 24 metres high. The angle of depression of both ends of a tennis court are 43° and 28° respectively.

a Calculate the length of the court.

b The net is half way along the court and is 1 metre high. What is the angle of depression of the top of the net from the building?

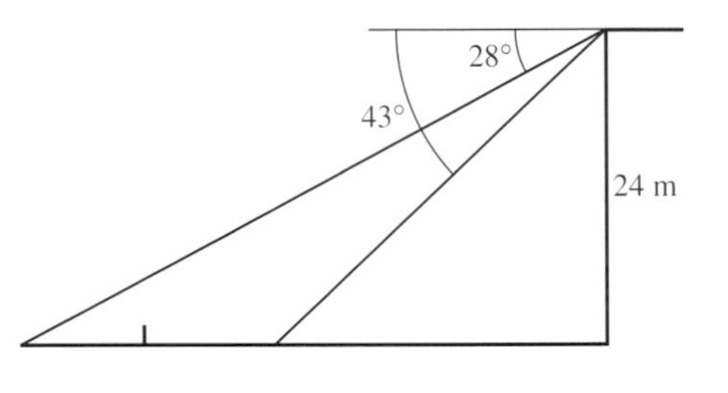

3 A ship leaves a port A and travels for 80 km on a bearing of 300° to a point B. It then turns and travels for 40 km on a bearing of 030° to a point C. Calculate

a how far west of A point C is **b** how far north of A point C is

c the bearing of A from C **d** the direct distance of A from C.

★4 A tower, CD, is at the top of a hill, BC. Martin measures the distance AC as 70 m and the angles of elevation of the top and bottom of the tower as 25° and 42° respectively. Calculate

a angle CAD **b** length AB

c length CB **d** height of tower, CD.

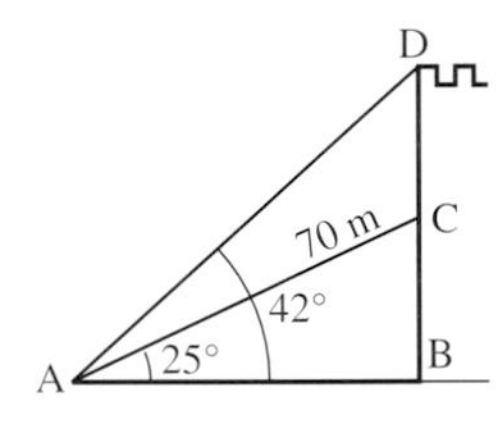

Homework 13B

1 The diagram shows a pyramid. The base is a square ABCD, 16 cm by 16 cm. The length of each sloping edge is 25 cm. The apex, V, is over the centre of the base. Calculate

a the size of the angle VAC

b the height of the pyramid

c the volume of the pyramid

d the size of the angle between the face VAD and the base ABCD.

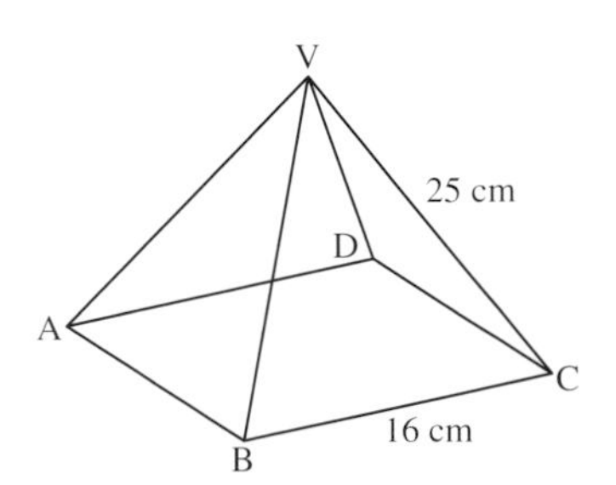

2 In the given cuboid, find

a angle AGE

b angle BMA

(M is the midpoint of GH)

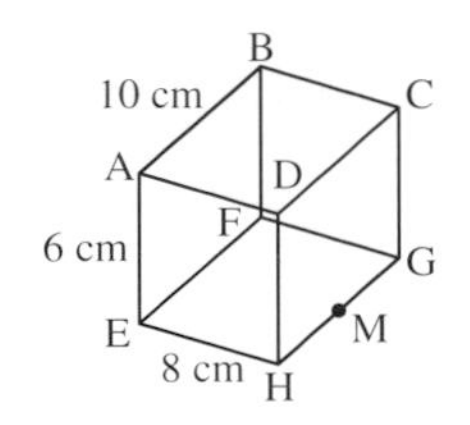

3 The diagram shows a wedge. Find

a CD

b angle CAD

c angle CAE

The point M is the midpoint of AB. Calculate

d the distance DM

e the angle of elevation of M from D.

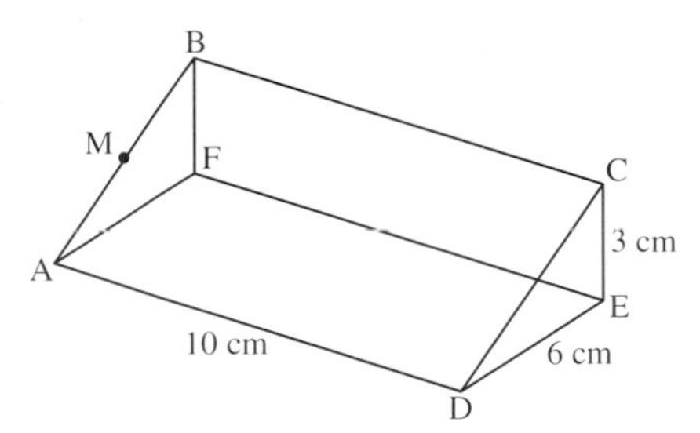

★4 A tetrahedron VPQR stands on a prism FGHPQR. The cross-section PQR is an equilateral triangle of side 8 cm. VP = VQ = VR = 10 cm. PF = QG = RH = 15 cm. M is the midpoint of QR.

a **i** Use triangle PQR to find the length of PM.

ii Use triangle VQR to find the length of VM.

b Find the size of angle VPM.

c Find the height of V above the base FGH. Give your answer to an appropriate degree of accuracy.

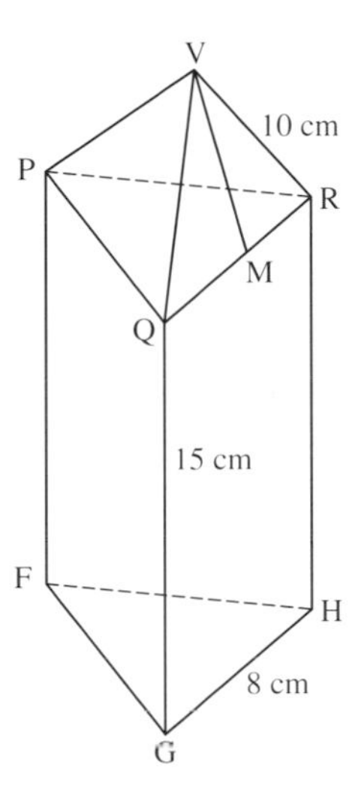

Homework 13C

State the two angles between 0° and 360° for each of these sine values.

1 0.4	**2** 0.45	**3** 0.65	**4** 0.27
5 0.453	**6** −0.4	**7** −0.15	**8** −0.52

9 Solve the equation $2 \sin x = 1$ for $0° \leq x \leq 360°$.

10 $\sin 40° = 0.643$. Write down the sine values of these angles.

i 140° **ii** 320° **iii** 400° **iv** 580°

★11 Solve the equation $3 \sin x = -2$ for $0° \leq x \leq 360°$.

Homework 13D

State the two angles between 0° and 360° for each of these cosine values.

1 0.7	**2** 0.38	**3** 0.617	**4** 0.376
5 0.085	**6** −0.6	**7** −0.45	**8** −0.223

9 Solve the equation $3 \cos x = -1$ for $0° \leq x \leq 360°$.

10 $\cos 50° = 0.643$. Write down the cosine values of these angles.
i 130° **ii** 310° **iii** 410° **iv** 590°.

★11 Solve the equation $6 \cos x = -1$ for $0° \leq x \leq 360°$.

HOMEWORK 13E

1 Write down the sine of each of these angles.
a 27° **b** 153° **c** 207° **d** 333°

2 Write down the cosine of each of these angles.
a 69° **b** 111° **c** 249° **d** 291°

3 What do you notice about the answers to Questions **1** and **2**?

4 Find four values between 0° and 360° such that
a $\sin x = \pm 0.4$ **b** $\cos x = \pm 0.5$

5 Solve **a** $\sin x + 1 = 2$ for $0° \leq x \leq 360°$ **b** $2 + 3 \cos x = 1$ for $0° \leq x \leq 360°$

★6 Find two values of x between 0° and 360° such that $\sin x = \cos 320°$.

HOMEWORK 13F

State the angles between 0° and 360° for each of these tangent values.

1 0.528 **2** 0.8 **3** 1.35 **4** 3.24
5 –2.55 **6** –0.158 **7** –0.786 **8** –1.999

9 Solve the equation $2 \tan x = -3$ for $0° \leq x \leq 360°$

10 $\tan 64° = 2.05$. Write down the tangent values of these angles.
i 116° **ii** 296° **iii** 424° **iv** 604°

★11 Solve the equation $2 \tan x - 5 = -1$ for $0° \leq x \leq 360°$.

HOMEWORK 13G

1 Find the length x in each of these triangles

a

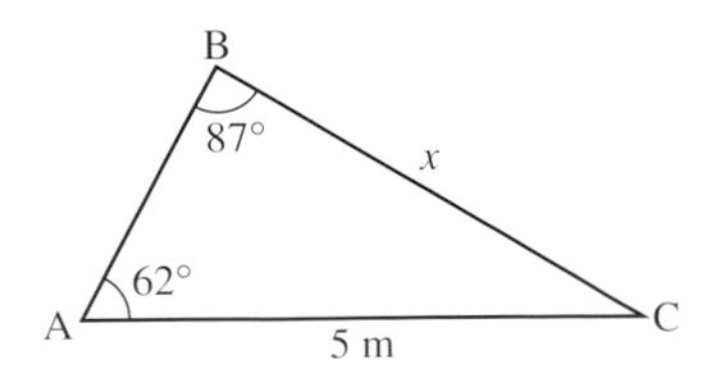

b

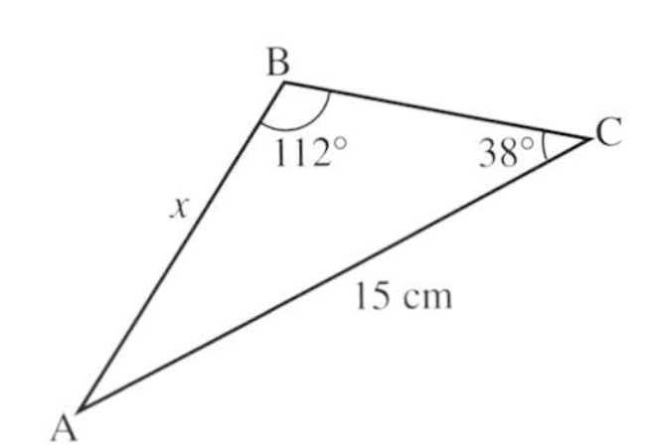

2 Find the angle x in each of these triangles

a

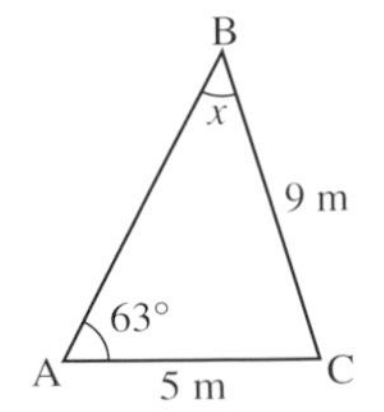

b

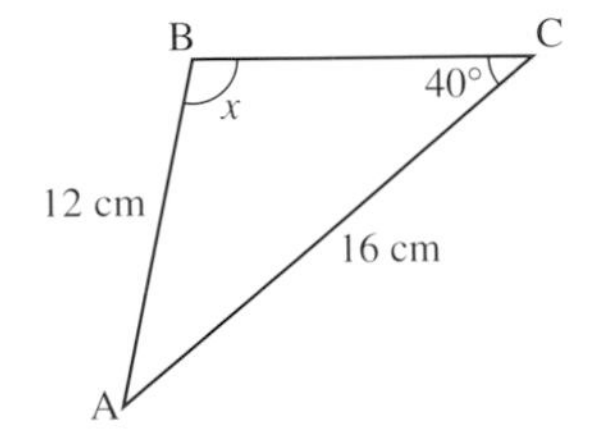

3 In triangle ABC, the angle at A is 40°, the side AB is 10 cm and the side BC is 7 cm. Find the two possible values of the angle at C.

4 In triangle ABC, the angle at A is 58°, the side AB is 20 cm and the side BC is 18 cm. Find the two possible values of the side AC.

5 To calculate the length of a submarine Mervyn stands on a cliff 60 m high and measures the angle of depression of both ends of the boat. The information is shown in the diagram.

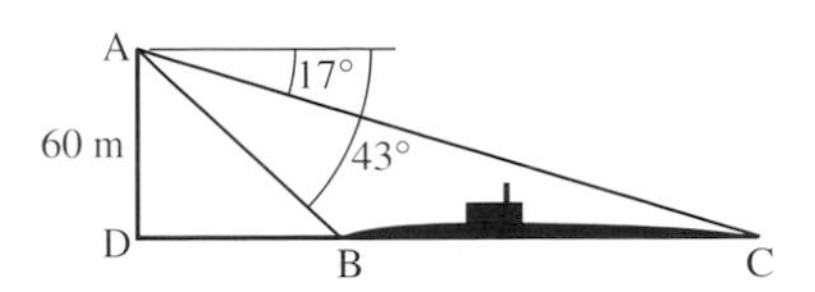

a Find the value of the angle DAB.

b Use trigonometry to calculate the length AB.

c Use the sine rule to find the length BC.

6 Use the information on this sketch to calculate the width, w, of the river.

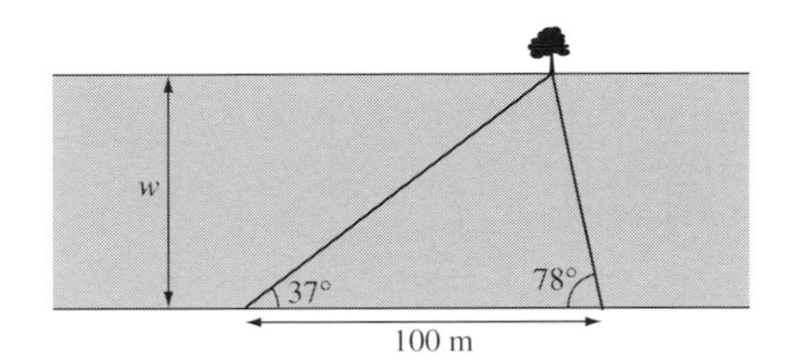

UAM **★7** A surveyor wishes to measure the height of a chimney. Measuring the angle of elevation, she finds that the angle increases from 28° to 37° after walking 30 metres towards the chimney. What is the height of the chimney?

HOMEWORK 13H

1 Find the length x in each of these triangles.

a

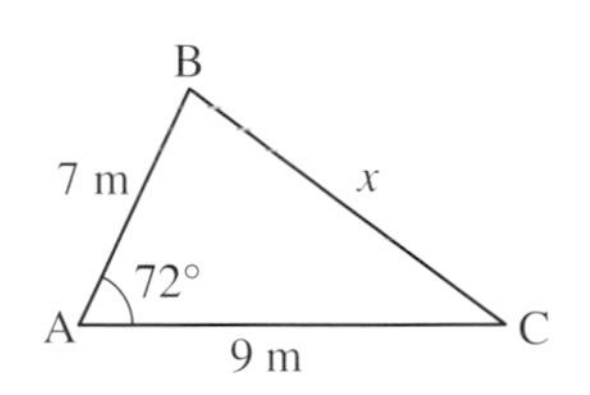

b

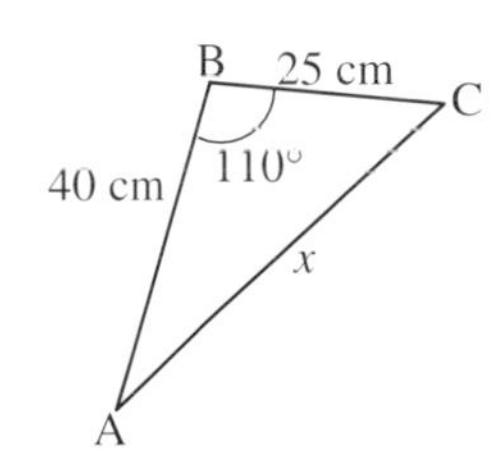

2 Find the angle x in each of these triangles.

a

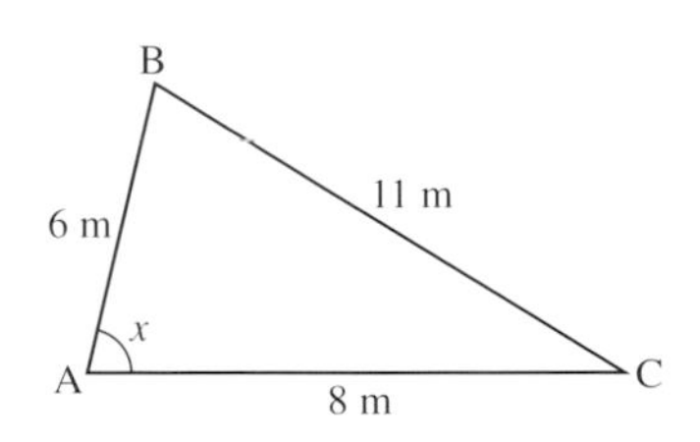

b

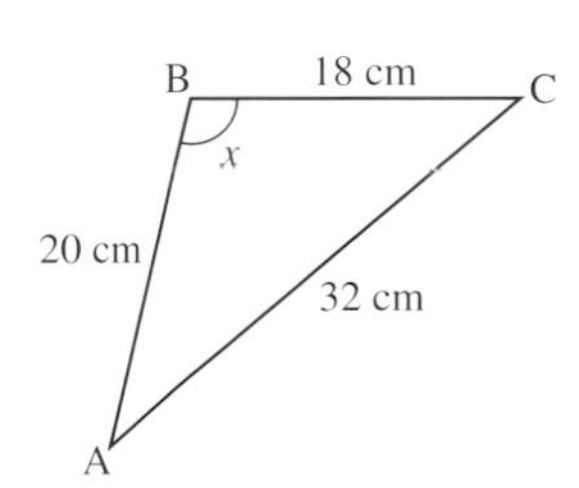

3 A quadrilateral ABCD has AD = 8 cm, DC = 10 cm, AB = 12 cm and BC = 15 cm. Angle ADC = 112°. Calculate angle ABC.

4 The three sides of a triangle are given as $2a$, $4a$ and $5a$. Calculate the smallest angle in the triangle.

5 The diagram shows a trapezium ABCD. AB = 6 cm, AD = 8 cm, CB = 12 cm and angle DAB = 115°. Calculate

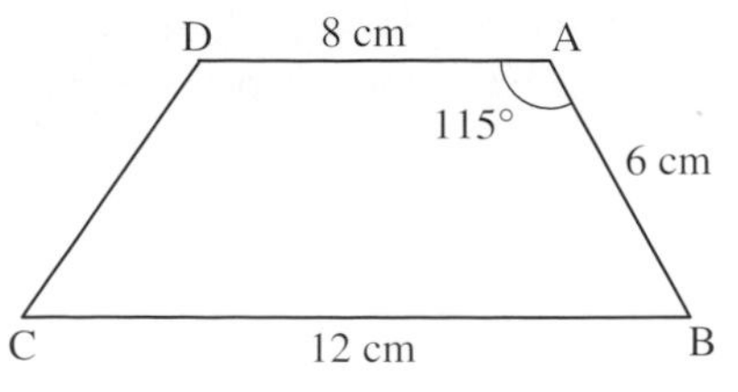

a length DB **b** angle DBA
c angle DBC **d** length DC
e area of the trapezium.

UAM ★**6** A port, B, is 20 km north east of another port, A. A lighthouse, L, is 5 km from B on a bearing of 260° from B. Calculate

a the distance AL **b** the bearing of L from A to the nearest degree.

HOMEWORK 13 I

1 Find the length or angle x in each of these triangles.

a

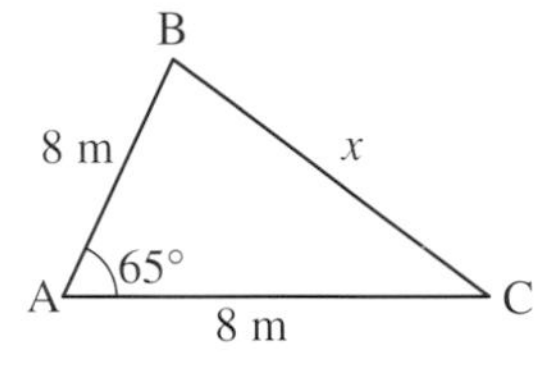

b

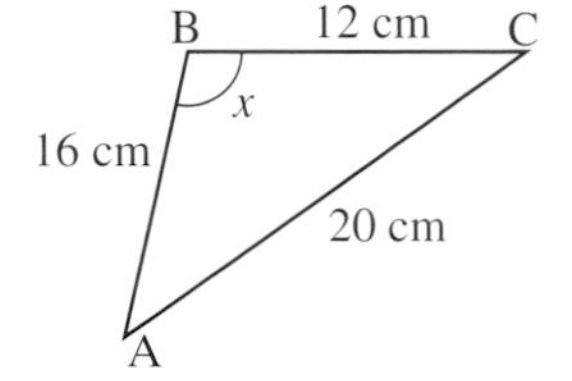

c

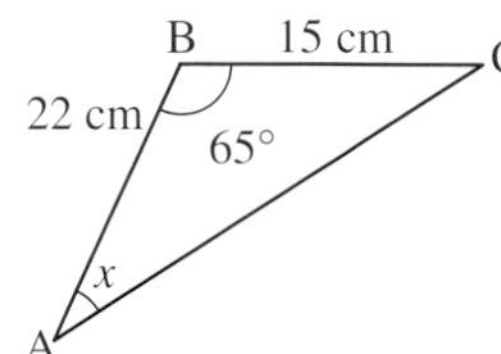

d

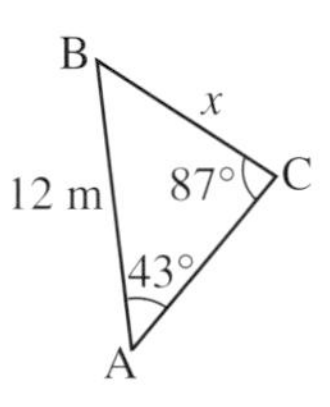

e

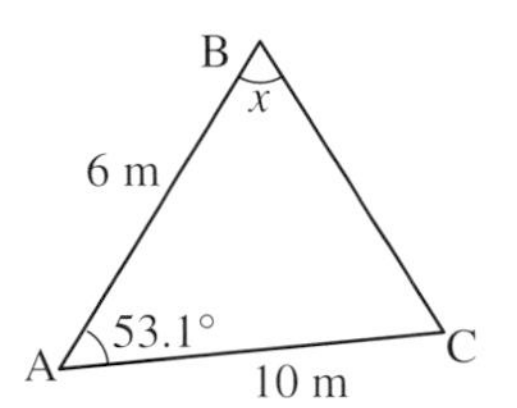

f

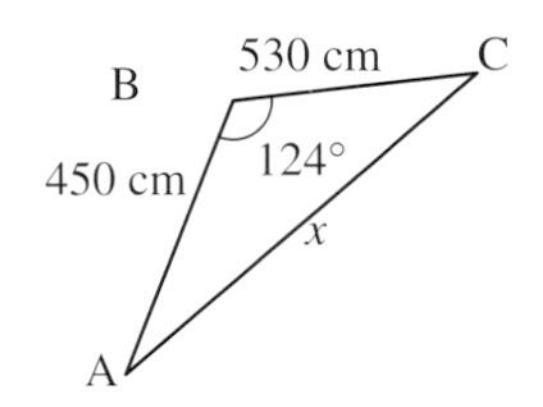

2 The hands of a clock have lengths 10 cm and 7 cm. Find the distance between the tips of the hands at 5 o'clock.

3 In the quadrilateral ABCD, find

a angle ABC
b the length of AC.

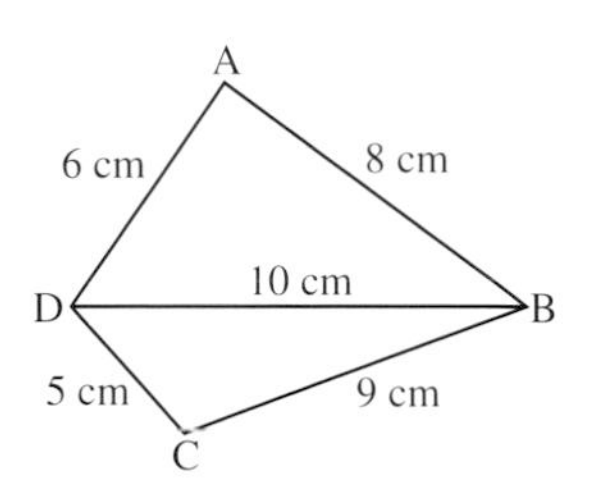

UAM ★**4** In a triangle, ABC, AC = 7.6 cm, angle BAC = 35°, angle ACB = 65°. The length of AB is x cm. Calculate the value of x.

HOMEWORK 13J

1 The sine of angle x is $\frac{3}{4}$.

Work out the cosine of angle x.

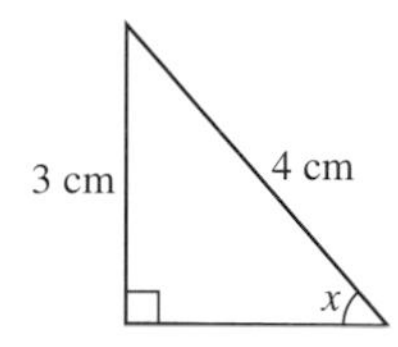

2 The cosine of angle x is $\frac{3}{\sqrt{18}}$.

Work out the value of angle x.

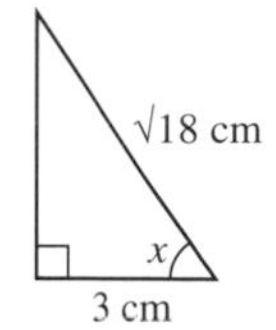

3 Calculate the exact value of the area of an equilateral triangle of side a cm.

4 Work out the exact value of the area of a right-angled isosceles triangle whose hypotenuse is 10 cm.

★5 A right-angled triangle has short sides of length $\sqrt{6}$ cm and $\sqrt{10}$ cm.

a Calculate the length of the hypotenuse.

b Write down the values of sin x and cos x.

c Show clearly that $(\sin x)^2 + (\cos x)^2 = 1$.

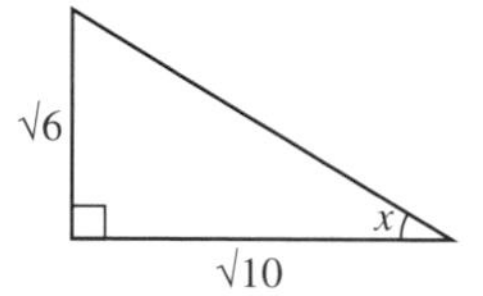

HOMEWORK 13K

1 Find the area of each of the following triangles.

a Triangle ABC where BC = 8 cm, AC = 10 cm and angle ACB = 69°.

b Triangle ABC where angle BAC = 112°, AC = 3 cm and AB = 7 cm.

2 The area of triangle ABC is 27 cm^2. If BC = 12 cm and angle BCA = 98°, find AC.

3 In a quadrilateral ABCD, DC = 3 cm, BD = 8 cm, angle BAD = 43°, angle ABD = 52° and angle BDC = 72°. Calculate the area of the quadrilateral.

4 The area of triangle LMN is 85 cm^2, LM = 10 cm and MN = 25 cm. Calculate

a angle LMN **b** angle MNL.

UAM **5** A board is in the shape of a triangle with sides 30 cm, 40 cm and 60 cm. Find its area.

UAM **★6** In the triangle ABC, angle B is **obtuse**, ∠BAC = 32°, AC = 10 cm, BC = 6 cm. Calculate the area of the triangle ABC.

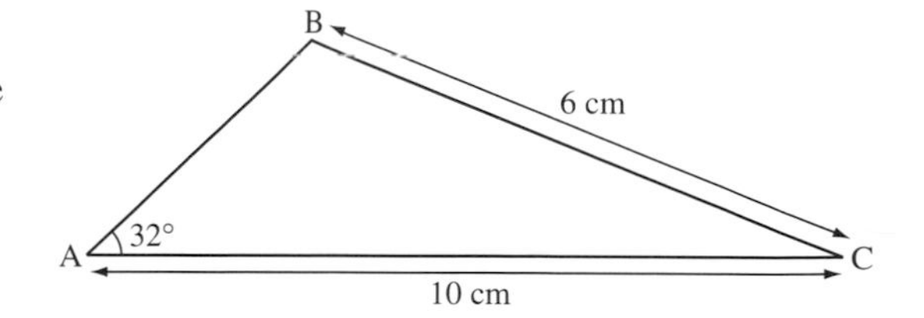

Chapter 14 Graphs and equations

Homework 14A

1 Give the equation of each of these lines.

a

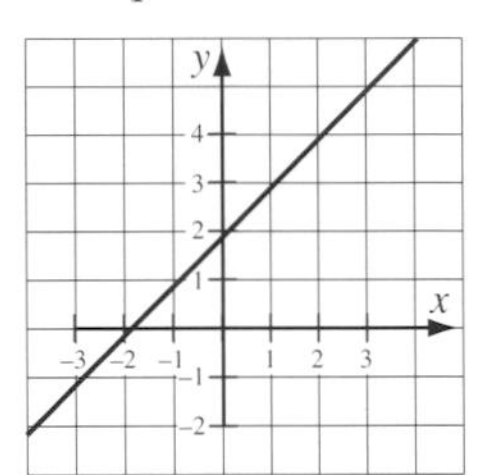

b

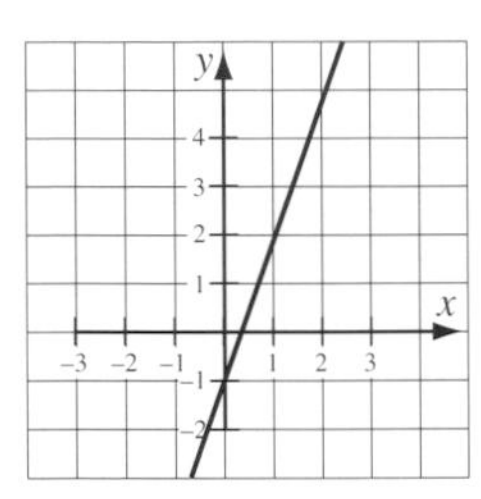

c

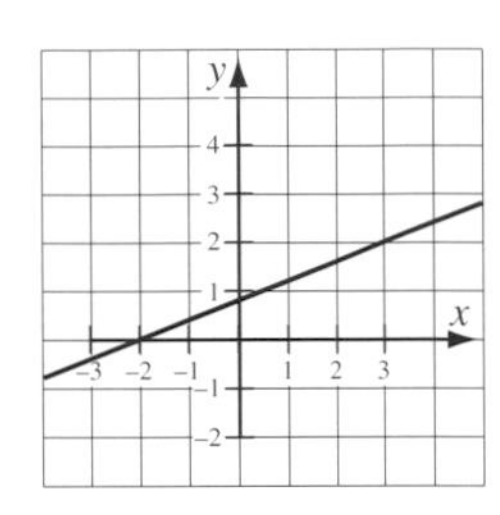

2 In each of these grids, there are two lines.

a

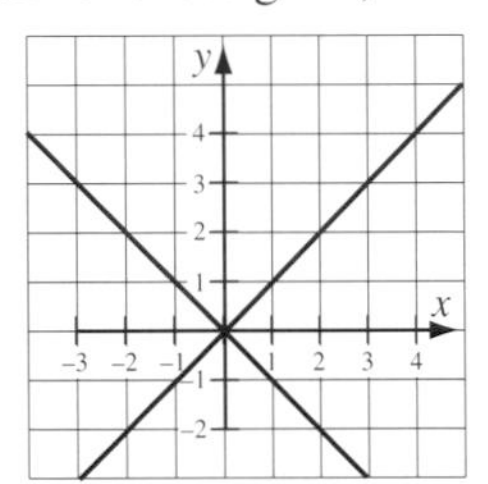

b

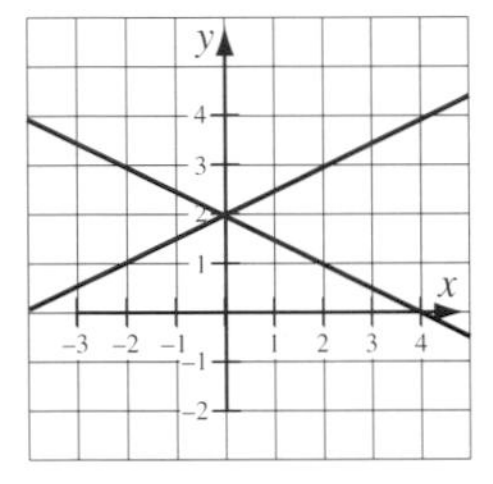

c

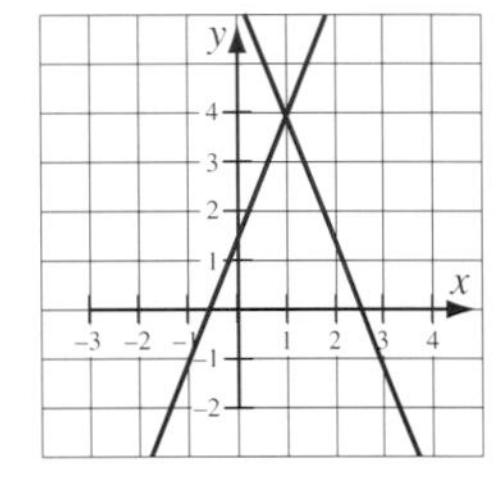

i Find the equation of each line.

ii Describe any symmetries you see about the two lines.

3 The diagram shows four lines crossing to create a rectangle. Write down the equation of each of the four lines.

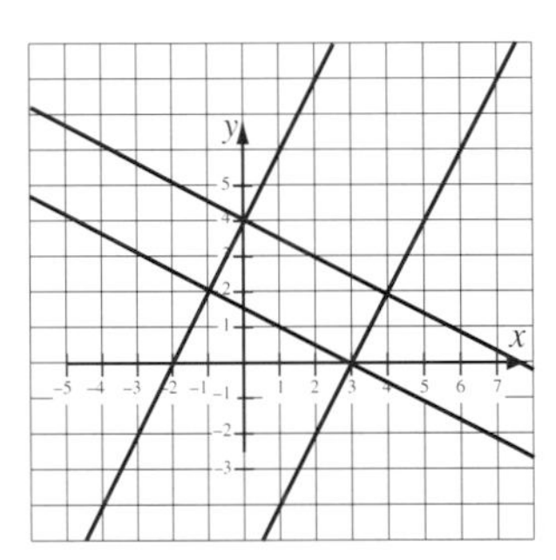

Homework 14B

1 Draw these lines using the cover-up method. Use the same grid, taking x from –10 to 10 and y from –10 to 10. If the grid gets too 'crowded', draw another.

a $2x + 3y = 6$ **b** $3x + 4y = 12$ **c** $5x - 4y = 20$ **d** $x + y = 8$

e $2x - 3y = 18$ **f** $x - y = 6$ **g** $3x - 5y = 15$ **h** $3x - 2y = 12$

i $5x + 4y = 30$ **j** $x + y = -1$ **k** $x + y = 5$ **l** $x - y = -6$

2 a Using the cover-up method, draw the following lines on the same grid.

i $x + 2y = 4$ **ii** $2x - y = 2$

b Where do the lines cross?

3 a Using the cover-up method, draw the following lines on the same grid.

i $x + 2y = 6$ **ii** $2x - y = 2$

b Where do the lines cross?

HOMEWORK 14C

1 This graph illustrates the charges made by an electricity company.

a Calculate the standing charge, this is the amount paid before any electricity is used.

b What is the gradient of the line?

c From your answers to **a** and **b** write down the rule to calculate the total charge for electricity.

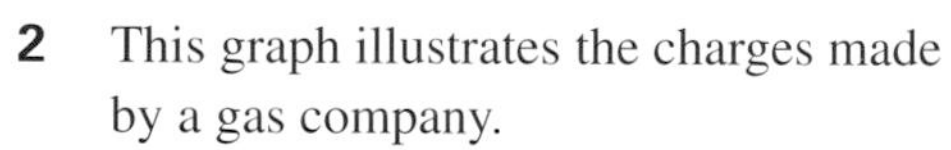

2 This graph illustrates the charges made by a gas company.

a Calculate the standing charge.

b What is the gradient of the line?

c From your answers to **a** and **b** write down the rule to calculate the total charge for gas.

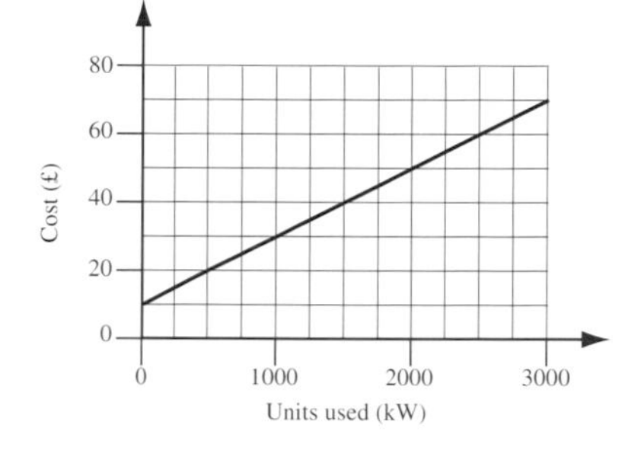

★3 This graph illustrates the charges made by a phone company.

a Calculate the standing charge.

b What is the gradient of the line?

c From your answers to **a** and **b** write down the rule to calculate the total charge for using this phone company.

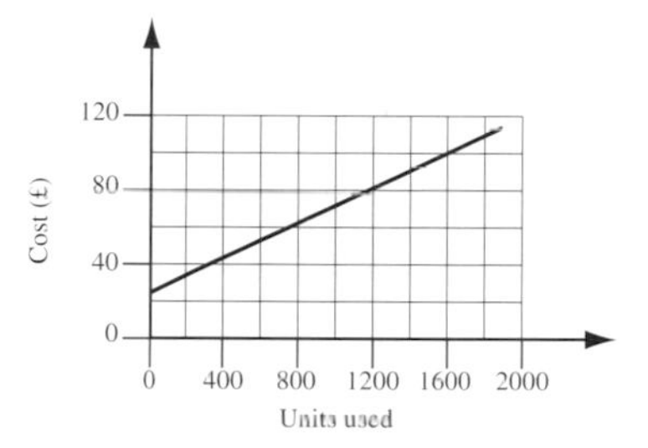

HOMEWORK 14D

By drawing their graphs, find the solution of each of these pairs of simultaneous equations.

1 $x + 4y = 1$ $x - y = 6$	**2** $y = 2x + 1$ $3x + 2y = 23$	**3** $y = 2x + 5$ $y = x + 4$	**4** $y = x$ $x - y = 4$
5 $y + 10 = 2x$ $5x + y = 18$	**6** $y = 5x - 1$ $y = 3x + 2$	**7** $y = x + 11$ $x + y = 5$	**8** $y - 3x = 8$ $y = x + 6$
9 $y = -x$ $y = 4x + 15$	**10** $3x + 2y = 2$ $y = -2x$	**11** $y = 3x - 4$ $y + x = 6$	**12** $y = 3x - 12$ $x + y = 2$

HOMEWORK 14E

1 **a** Copy and complete the table to draw the graph of $y = 2x^2$ for $-3 \leq x \leq 3$.

x	-3	-2	-1	0	1	2	3
$y = 2x^2$	18		2			8	

b Use your graph to find the value of y when $x = -1.4$.

c Use your graph to find the values of x that give a y-value of 10.

2 **a** Copy and complete the table to draw the graph of $y = x^2 + 3$ for $-5 \leq x \leq 5$.

x	-5	-4	-3	-2	-1	0	1	2	3	4	5
$y = x^2 + 3$	28		12					7			28

b Use your graph to find the value of y when $x = 2.5$.

c Use your graph to find the values of x that give a y-value of 10.

★3 **a** Copy and complete the table to draw the graph of $y = x^2 - 3x + 2$ for $-3 \leq x \leq 4$.

x	-3	-2	-1	0	1	2	3	4
$y = x^2 - 3x + 2$	20			2			2	

b Use your graph to find the value of y when $x = -1.5$.

c Use your graph to find the values of x that give a y-value of 2.5.

HOMEWORK 14F

1 **a** Copy and complete the table to draw the graph of $y = x^2 - 3x + 2$ for $-1 \leq x \leq 5$.

x	-1	0	1	2	3	4	5
$y = x^2 - 3x + 2$	6	2			1		

b Use your graph to find the roots of the equation $x^2 - 3x + 2 = 0$

2 **a** Copy and complete the table to draw the graph of $y = x^2 - 5x + 4$ for $-1 \leq x \leq 6$.

x	-1	0	1	2	3	4	5	6
$y = x^2 - 5x + 4$	10	4				0		

b Use your graph to find the roots of the equation $x^2 - 5x + 4 = 0$

3 **a** Copy and complete the table to draw the graph of $y = x^2 + 4x - 6$ for $-5 \leq x \leq 2$.

x	-5	-4	-3	-2	-1	0	1	2
$y = x^2 + 4x - 6$	-1							6

b Use your graph to find the roots of the equation $x^2 + 4x - 6 = 0$

HOMEWORK 14G

1 **a** Complete the table to draw the graph of $y = \frac{12}{x}$ for $-12 \le x \le 12$.

x	−12	−6	−4	−3	−2	−1	1	2	3	4	6	12
$y = \frac{12}{x}$	−1			−4					4			1

b Use your graph to find

i the y-value when $x = 1.5$ **ii** the x-value when $y = 5.5$

★**2** **a** Complete the table to draw the graph of $y = \frac{8}{x}$ for $-8 \le x \le 8$.

x	−8	−5	−4	−2	−1	1	2	4	5	8
$y = \frac{8}{x}$										

b Use your graph to find

i the y-value when $x = 3.5$ **ii** the x-value when $y = 5$

3 **a** Complete the table to draw the graph of $y = \frac{50}{x}$ for $0 \le x \le 50$.

x	1	2	5	10	25	50
$y = \frac{50}{x}$						

b On the same axes, draw the line $y = x + 30$.

c Use your graph to find the x-values of the points where the graphs cross.

HOMEWORK 14H

1 **a** Complete the table to draw the graph of $y = x^3 + 1$ for $-3 \le x \le 3$.

x	−3	−2	−1	0	1	2	3
$y = x^3 + 1$	−26			1			28

b Use your graph to find the y-value for an x-value of 1.2.

★**2** **a** Complete the table to draw the graph of $y = x^3 + 2x$ for $-2 \le x \le 3$.

x	−2	−1	0	1	2	3
$y = x^3 + 2x$	−12		0		12	

b Use your graph to find the y-value for an x-value of 2.5.

3 **a** Draw the graph of $y = x^3 - x^2$ for $-3 \le x \le 3$.

b Use your graph to find the y-value for an x-value of 1.8.

HOMEWORK 14I

1 **a** Complete the table below for $y = 2^x$ for values of x from –3 to +4. (Values are rounded to 2dp.)

x	–3	–2	–1	0	1	2	3	4
$y = 2^x$	0.1	0.3			2	4		

b Plot the graph of $y = 2^x$ for $-3 \leq x \leq 4$. (Take y-axis from 0 to 20.)

c Use your graph to estimate the value of y when $x = 2.5$.

d Use your graph to estimate the value of x when $y = 0.75$.

2 **a** Complete the table below for $y = (\frac{1}{3})^x$ for values of x from –3 to +3. (Values are rounded to 2dp.)

x	–3	–2	–1	0	1	2	3
$y = (\frac{1}{3})^x$	27			1			0.04

b Plot the graph of $y = (\frac{1}{3})^x$ for $-3 \leq x \leq 3$. (Take y axis from 0 to 30.)

c Use your graph to estimate the value of y when $x = 2.5$.

d Use your graph to estimate the value of x when $y = 0.75$.

3 Granny has two nephews, Alf and Bert. She writes a will leaving Alf £1000 in the first year after her death, £2000 in the second year after her death, £3000 the next year and so on for 20 years. She leaves Bert £1 the first year, £2 the second year, £4 the next year and so on.

a Show that the formula $500n(n + 1)$ gives the total amount that Alf gets after n years.

b Show that the formula $2^n - 1$ gives the total amount that Bert gets after n years.

c Complete the table for the total amount of money that Bert gets.

Year	2	4	6	8	10	12	14	16	18	20
Total	3	15	63							

d Draw a graph of both nephews' total over 20 years. Take the x-axis from 0 to 20 years and the y-axis from £0 to £1100 000.

e Which nephew gets the best deal?

HOMEWORK 14J

1 Below is the graph of $y = x^2 - 2x - 4$. Use this graph to solve

a $x^2 - 2x - 4 = 0$

b $x^2 - 2x - 4 = 4$

c $x^2 - 2x - 3 = 0$

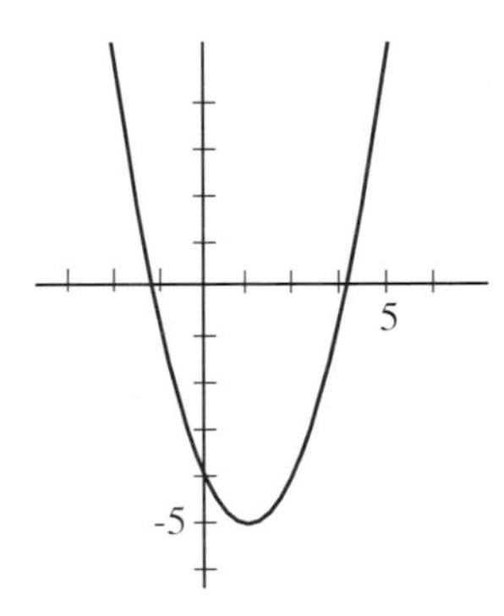

2 Below are the graphs of $y = x^2 - 3x + 1$ and $y = x - 1$. Use these graphs to solve

a $x^2 - 3x + 1 = 0$

b $x^2 - 4x + 2 = 0$

c What straight-line graph will you need to draw to solve $x^2 - 4x + 3$? Draw this line and write down the two solutions to $x^2 - 4x + 3$.

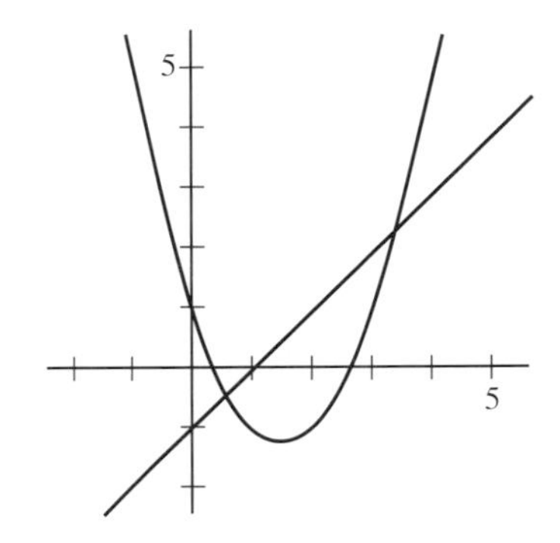

3 Draw the graph of $y = x^3 - 2x + 3$.

a Use the graph to solve **i** $x^3 - 2x + 3 = 0$ **ii** $x^3 - 2x = 0$

b What straight-line graph will you need to draw to solve $x^3 - 3x + 2 = 0$? Draw this line and solve $x^3 - 3x + 2 = 0$.

4 The graph of $y = x^3 - 4x - 1$ is shown on the right.

a Use the graph to solve

i $x^3 - 4x - 1 = 0$

ii $x^3 - 4x + 2 = 0$

b By drawing an appropriate straight line solve the equation $x^3 - 5x - 1 = 0$.

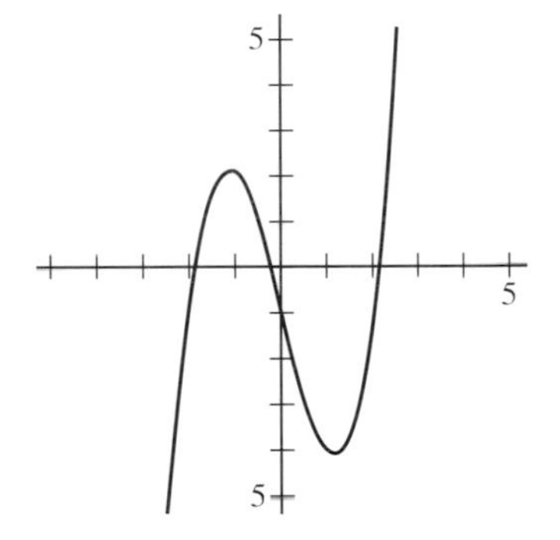

★5 The graph of $y = x^3 - 4x$ is shown.

a Use the graph to find the two positive solutions to $x^3 - 4x = -2$.

b By drawing an appropriate straight line use the graph to solve $x^3 - 3x + 1 = 0$.

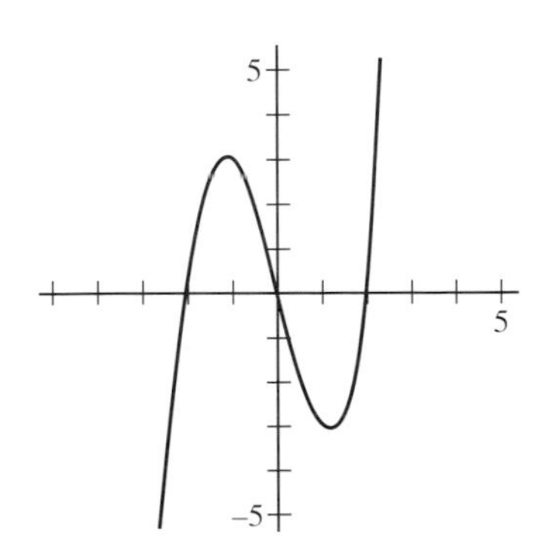

HOMEWORK 15A

1 The table shows the heights and weights of twelve students in a class.

a Plot the data on a scatter diagram.

b Draw the line of best fit.

c Jayne was absent from the class, but they knew she was 132 cm tall. Use the line of best fit to estimate her weight.

d A new girl joined the class who weighed 55 kg. What height would you expect her to be?

Student	Weight (kg)	Height (cm)
Ann	51	123
Bridie	58	125
Ciri	57.5	127
Di	62	128
Emma	59.5	129
Flo	65	129
Gill	65	133
Hanna	65.5	135
Ivy	71	137
Joy	75.5	140
Keri	70	143
Laura	78	145

★2 The table shows the marks for ten pupils in their mathematics and music examinations.

a Plot the data on a scatter diagram. Take the x-axis for the mathematics scores and mark it from 20 to 100. Take the y-axis for the music scores and mark it from 20 to 100.

b Draw the line of best fit.

c One of the pupils was ill when they took the music examination. Which pupil was it most likely to be?

d Another pupil, Kris, was absent for the music examination but scored 45 in mathematics, what mark would you expect him to have got in music?

e Another pupil, Lex, was absent for the mathematics examination but scored 78 in music, what mark would you expect him to have got in mathematics?

Pupil	Maths	Music
Alex	52	50
Ben	42	52
Chris	65	60
Don	60	59
Ellie	77	61
Fan	83	74
Gary	78	64
Hazel	87	68
Irene	29	26
Jez	53	45

HOMEWORK 15B

1 An army squad were all sent on a one mile run. Their coach recorded the times they actually took. This table shows the results.

a Copy the table and complete a cumulative frequency column.

b Draw a cumulative frequency diagram.

c Use your diagram to estimate the median time and the interquartile range.

Time (seconds)	Number of runners
$200 < x \le 240$	3
$240 < x \le 260$	7
$260 < x \le 280$	12
$280 < x \le 300$	23
$300 < x \le 320$	7
$320 < x \le 340$	5
$340 < x \le 360$	5

★2 A company had some web pages. They recorded how many times they were visited on one day.

a Copy the table and complete a cumulative frequency column.

b Draw a cumulative frequency diagram.

c Use your diagram to estimate the median use of the web pages and the interquartile range.

d Pages with less than 60 visitors are going to be rewritten. About how many pages would need to be rewritten?

Number of visits	Number of pages
$0 < x \le 50$	6
$50 < x \le 100$	9
$100 < x \le 150$	15
$150 < x \le 200$	25
$200 < x \le 250$	31
$250 < x \le 300$	37
$300 < x \le 350$	32
$350 < x \le 400$	17
$400 < x \le 450$	5

HOMEWORK 15C

1 The box plot below shows the number of peas in pods grown by a prize-winning gardener.

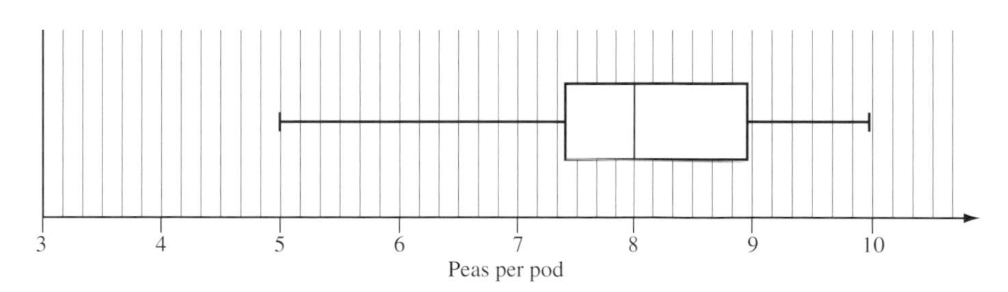

A young gardener also grew some peas. These are the results for the number of peas per pod. Smallest number 3, Lower quartile 4.75, Median 5.5, Upper quartile 6.25, Biggest number 9.

a Copy the diagram and draw a box plot for the young gardener.

b Comment on the differences between the two distributions.

2 The box plot shows the monthly salaries of the men in a computer firm.

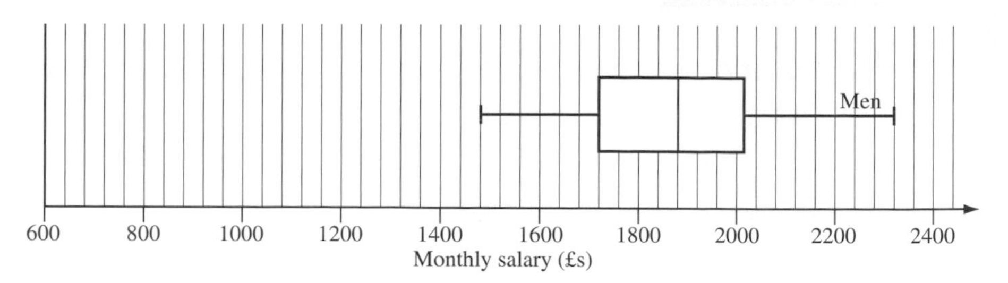

The data for the women in the company is Smallest 600, Lower quartile 1300, Median 1600, Upper Quartile 2000, Largest 2400.

a Copy the diagram and draw a box plot for the women's salaries.

b Comment on the differences between the two distributions.

3 The box plots for the hours of life of two brands of batteries are shown below.

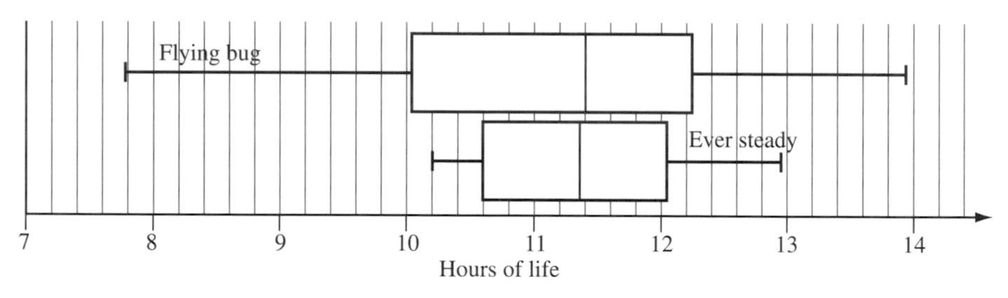

a Comment on the differences in the two distributions.

b Mushtaq wants to get some batteries for his palm top. Which brand would you recommend and why?

4 The following table shows some data on the times of telephone calls to two operators at a mobile phone helpline.

	Lowest time	Lowest Quartile	Medium time	Upper Quartile	Highest time
Jack	1 m 10 s	2 m 20 s	3 m 30 s	4 m 50 s	7 m 10 s
Jill	40 s	2 m 20 s	5 m 10 s	7 m 30 s	10 m 45 s

a Draw box plots to compare both sets of data.

b Comment on the differences between the distributions.

c The company has to get rid of 1 operator. Who should they get rid of and why?

★5 A school entered 80 pupils for an examination. The results are shown in the table.

Mark, x	$0 < x \leq 20$	$20 < x \leq 40$	$40 < x \leq 60$	$60 < x \leq 80$	$80 < x \leq 100$
Number of pupils	2	14	28	26	10

a Calculate an estimate of the mean.

b Complete a cumulative frequency table and draw a cumulative frequency diagram.

c **i** Use your graph to estimate the median mark.

ii 12 of these pupils were given a grade A. Use your graph to estimate the lowest mark for which grade A was given.

d Another school also entered 80 pupils for the same examination. Their results were Lowest mark 40, Lower quartile 50, Median 60, Upper quartile 70, Highest mark 80. Draw a box plot to show these results and use it to comment on the differences between the two schools' results.

HOMEWORK 15D

Use your calculator to work out the answers.

1 Find the mean and standard deviation of the following sets of data.

a 4, 7, 8, 8, 11 **b** 17, 19, 21, 23, 28 **c** 82, 85, 88, 89, 92
d 201, 202, 203, 204, 205 **e** 65, 71, 76, 76 **f** −4, −2, −1, 0, 2, 5, 7
g 91, 92, 98, 100, 105 **h** 29, 32, 35, 36, 42, 43

2 **a** Calculate the mean and the standard deviation of 6, 8, 11, 14, 15, 18.
b Write down the mean and standard deviation of 16, 18, 21, 24, 25, 28.
c Write down the mean and the standard deviation of 60, 80, 110, 140, 150, 180.

3 Five numbers, 4, 8, 10, x and y, have a mean of 10 and a standard deviation of 4. Find the values of x and y.

4 In the 1996 Northern Counties Road Relays, teams of six runners each ran a 4-mile leg. The three winning teams had these times for each of their six runners.

First team: Salford Harriers 19.50, 19.47, 19.43, 19.30, 20.01, 19.40
Second team: Bingley Harriers 19.42, 20.32, 20.09, 19.50, 19.34, 19.36
Third team: Sunderland 20.40, 20.40, 20.36, 19.23, 20.29, 19.45

a Calculate the mean and the standard deviation of each team. (Remember that the times above are in minutes and seconds.)
b One of the top three teams has to be chosen to represent Northern Counties at the National Road Relays. Which team should be chosen and why?

HOMEWORK 15E

Use your calculator to work out the answers.

1 The number of eggs in 20 nests in a wood are

3, 6, 5, 8, 3, 4, 5, 7, 7, 8, 3, 9, 3, 6, 2, 6, 5, 8, 7, 2

Put these numbers into a frequency table and calculate the mean and the standard deviation of the number of eggs per nest.

2 Calculate the mean and the standard deviation of the data in these tables.

a

x	f
0	3
1	5
2	9
3	5
4	3

b

x	f
122	4
123	6
124	7
125	6
126	4
127	3

c

x	f
100	3
101	5
102	9
103	5
104	3

3 The tables below show the results of two classes in a mathematics test.

Class 1

Mark, x	0	1	2	3	4	5	6	7	8	9	10
Frequency, f	0	1	4	6	5	4	5	6	1	2	1

Class 2

Mark, x	0	1	2	3	4	5	6	7	8	9	10
Frequency, f	0	0	2	2	10	11	5	4	1	1	0

a Calculate the mean and the standard deviation for each class.
b Which class did better in the test? Why?

4 During the first six weeks in a new job Mrs Best recorded how long it took her from getting out of bed to walking through the office door. The results are shown in the table.

Time, x (min)	80	85	90	95	100	105
Frequency, f	6	4	10	5	3	2

a Calculate the mean and the standard deviation of the data.
b Mrs Best has to arrive at work by 9.00 am. If she is late more than five times in a 30-day period, she will be sacked. She decides to set her alarm for 7.30 am. Is this a good idea? Justify your answer.

Chapter 16 Probability

HOMEWORK 16A

1 Kylie takes a ball at random from a bag that contains 6 red and 4 white balls. The table shows her results.

Number of draws	10	20	50	100	500
Number of white balls	2	6	18	42	192

a Calculate the relative frequency of a white ball at each stage that Kylie recorded her results.
b What is the theoretical probability of taking a white ball from the bag?
c If Kylie took a ball out of the bag a total of 5000 times, how many white balls would you expect her to get?

2 Jason made a six-sided spinner. To test it he threw it 600 times. The table shows the results.

Number	1	2	3	4	5	6
Total	98	152	85	102	62	101

a Work out the relative frequency of each number.
b How many times would you expect each number to occur if the spinner is fair?
c Do you think that the spinner is fair? Give a reason for your answer.

3 A bag contains red, white and blue balls. It is known that there are 50 balls in the bag altogether. Mary performs an experiment to see how many of each colour are in the bag. She counts up the tallies and records them in a table.

Red	White	Blue	Total
31	52	17	100
68	120	62	250
102	203	95	400
127	252	121	500

a Calculate the relative frequencies of the balls at each stage to 3 sf.

b How many of each colour do you think are in the bottle? Explain your answer.

4 Which of these methods would you use to estimate or state the probability of each of the events **a** to **g**?

Method A: Equally likely outcomes   Method B: Survey or experiment
Method C: Look at historical data

a There will be an earthquake in Japan.

b The next person to walk through the door will be female.

c A Premier League team will win the FA Cup.

d You will win a raffle.

e The next car to drive down the road will be foreign.

f You will have a Maths lesson this week.

g A person picked at random from your school will go abroad for their holiday.

★5 A bag contains a number of counters. Each counter is coloured red, blue or yellow. Each counter is numbered 1 or 2. The table shows the probability of the colour and number for those counters.

	Colour of counter		
Number of counter	Red	Blue	Yellow
1	0.2	0.3	0.1
2	0.2	0.1	0.1

a A counter is taken from the bag at random. What is the probability that

i it is red **and** numbered 2   **ii** it is blue **or** numbered 2

iii it is red **or** numbered 2?

b There are two yellow counters in the bag. How many counters are in the bag altogether?

Homework 16B

1 What is the probability of each of the following?

a Throwing a 3 with a dice.   **b** Tossing a coin and getting a head.

c Drawing a Jack from a pack of cards.   **d** Drawing a Club from a pack of cards.

e Throwing a 3 or a 6 with a dice.

2 What is the probability of each of the following?

a Throwing an odd number with a dice.

b Throwing a square number with a dice.

c Getting a Spade or a Diamond from a pack of cards.

d Drawing the Jack of Diamonds from a pack of cards.

3 A bag contains 3 blue balls, 2 pink balls and 1 black ball. Joan takes a ball from the bag without looking. What is the probability that she takes out

a a blue ball **b** a pink ball **c** a ball that is not black?

4 In a raffle, 800 tickets are sold. Dick has 5 tickets. What is the probability that he wins the first prize?

5 Ann, Betty, Colin, Derek, Ethel and Fiona are in the same class. Their teacher wants two pupils to do a special job.

a Write down all the possible combinations of two people. For example, Ann and Betty, Ann and Colin (there are 15 combinations altogether).

b How many pairs give two boys?

c What is the probability of choosing two boys?

d How many pairs give a boy and a girl?

e What is the probability of choosing a boy and a girl?

f What is the probability of choosing two girls?

6 A bag contains 12 identical tea bags. Five are Earl Grey tea and the rest are ordinary tea. I take one out and make a cup of tea with it

a What is the probability that I get Earl Grey tea?

b If the first tea bag I get is Earl Grey,

i how many Earl Grey tea bags are left

ii how many ordinary tea bags are left?

c After I have drunk the cup of Earl Grey tea, I pick another tea bag. What is the probability that I pick

i Earl Grey **ii** ordinary tea?

7 An ordinary six-sided dice has 1 red face, 1 blue face and 4 green faces. If this dice is thrown, what is the probability that the top face will be

a red **b** green **c** not blue

d black **e** red, green or blue?

8 Twelve-sided dice are used in adventure games. They are marked with the numbers 1 to 12. The score is the uppermost face. If a twelve-sided dice is thrown, what is the probability that the score will be

a a number in the 3 times table **b** a factor of 12

c a square number **d** a triangle number

e a number that is not prime **f** not a square number?

★9 Zaheda conducted a probability experiment using a packet of 40 sweets. She counted the number of sweets of each colour. Her results are shown in the table. Zaheda is going to take one sweet at random from the packet. Write down the probability

Red	Green	Orange
16	9	15

a that Zaheda will take a green sweet from the packet

b that the sweet Zaheda takes will **not** be red.

HOMEWORK 16C

1 Say whether these pairs of events are mutually exclusive or not.

a Tossing two heads with two coins/tossing two tails with two coins.

b Throwing an even number with a dice/throwing an odd number with a dice.

c Drawing a Queen from a pack of cards/drawing an Ace from a pack of cards.

d Drawing a Queen from a pack of cards/drawing a red card from a pack of cards.

e Drawing a red card from a pack of cards/drawing a heart from a pack of cards.

2 Which of the pairs of mutually exclusive events in Question **1** are also exhaustive?

3 A letter is to be chosen at random from this set of letter-cards.
M I S S I S S I P P I

a What is the probability the letter is
i an S **ii** a P **iii** a vowel?

b Which of these pairs of events are mutually exclusive?
i Picking an S / picking a P. **ii** Picking an S / picking an I.
iii Picking an I / picking a consonant.

c Which pair of mutually exclusive events in part **b** is also exhaustive?

4 Two people are to be chosen for a job from this set of six people.
Ann Joan Jack John Arthur Ethel

a List all of the possible pairs (there are 15 altogether).

b What is the probability that the pair of people chosen will be
i both female **ii** both male **iii** both have the same initial
iv have different initials?

c Which of these pairs of events are mutually exclusive?
i Picking two women / picking two men.
ii Picking two people of the same sex / picking two people of opposite sex.
iii Picking two people with the same initial / picking two men.
iv Picking two people with the same initial / picking two women.

d Which pair of mutually exclusive events in part **c** is also exhaustive?

5 For breakfast I like to have toast, porridge or cereal. The probability I have toast is $\frac{1}{3}$, the probability I have porridge is $\frac{1}{2}$, what is the probability I have cereal?

6 A person is chosen at random. Here is a list of events.
Event A: the person chosen is male Event B: the person chosen is female
Event C: the person chosen is over 18 Event D: the person chosen is under 16
Event E: the person chosen has a degree Event F: the person chosen is a teacher

For each of the pairs of events **i** to **x**, say whether they are

a mutually exclusive **b** exhaustive.

c If they are not mutually exclusive, explain why.

i Event A and Event B **ii** Event A and Event C
iii Event B and Event D **iv** Event C and Event D
v Event D and Event F **vi** Event E and Event F
vii Event E and Event D **viii** Event A and Event E
ix Event C and Event F **x** Event C and Event E

★7 An amateur weather man records the weather over a year in his village. He knows that the probability of a windy day is 0.4 and that the probability of a rainy day is 0.6. Steve says 'This means it will be either rainy or windy each day as 0.4 + 0.6 = 1, which is certain.' Explain why Steve is wrong.

HOMEWORK 16D

1 I throw an ordinary dice 600 times. How many times can I expect to get a score of 1?

2 I toss a coin 500 times. How many times can I expect to get a tail?

3 I draw a card from a pack of cards and replace it. I do this 104 times. How many times would I expect to get
 a a red card **b** a Queen **c** a red seven
 d the Jack of Diamonds?

4 The ball in a roulette wheel can land on any number from 0 to 36. I always bet on the same block of numbers 0–6. If I play all evening and there is a total of 111 spins of the wheel in that time, how many times could I expect to win?

5 I have 5 tickets for a raffle and I know that the probability of my winning the prize is 0.003. How many tickets were sold altogether in the raffle?

6 In a bag there are 20 balls, 10 of which are red, 3 yellow, and 7 blue. A ball is taken out at random and then replaced. This is repeated 200 times. How many times would I expect to get
 a a red ball **b** a yellow or blue ball
 c a ball that is not blue **d** a green ball?

7 A sampling bottle contains black and white balls. It is known that the probability of getting a black ball is 0.4. How many white balls would you expect to get in 200 samples?

8 **a** Fred is about to take his driving test. The chance he passes is $\frac{1}{3}$. His sister says 'Don't worry if you fail because you are sure to pass within three attempts because $3 \times \frac{1}{3} = 1$'. Explain why his sister is wrong.
 b If Fred does fail would you expect the chance that he passes next time to increase or decrease? Explain your answer.

★9 An opinion poll used a sample of 200 voters in one area. 112 said they would vote for Party A. There are a total of 50 000 voters in the area.
 a If they all voted, how many would you expect to vote for Party A?
 b The poll is accurate within 10%. Can Party A be confident of winning?

HOMEWORK 16E

1 Shaheeb throws an ordinary dice. What is the probability that he throws
 a an even number **b** 5 **c** an even number or 5?

2 Jane draws a card from a pack of cards. What is the probability that she draws
 a a red card **b** a black card **c** a red or a black card?

3 Natalie draws a card from a pack of cards. What is the probability that she draws one of the following numbers?
 a Ace **b** King **c** Ace or King

4 A letter is chosen at random from the letters in the word STATISTICS. What is the probability that the letter will be
 a S **b** a vowel **c** S or a vowel?

5 A bag contains 10 white balls, 12 black balls and 8 red balls. A ball is drawn at random from the bag. What is the probability that it will be

a white **b** black **c** black or white
d not red **e** not red or black?

6 A spinner has numbers and colours on it, as shown in the diagram. Their probabilities are given in the tables.

Red	0.5
Green	0.25
Blue	0.25

1	0.4
2	0.35
3	0.25

When the spinner is spun what is the probability of each of the following?

a Red or blue **b** 2 or 3 **c** 3 or blue **d** 2 or green
e **i** Explain why the answer to **c** is 0.25 and not 0.5.
ii What is the answer to P(2 or red)?

7 Debbie has 10 CDs in her multi-changer, 4 of which are rock, 2 are dance and 4 are classical. She puts the player on random play. What is the probability that the first CD will be

a rock or dance **b** rock or classical **c** not rock?

8 Frank buys 1 dozen free-range eggs. The farmer tells him that a quarter of the eggs his hens lay have double yolks.

a How many eggs with double yolks can frank expect to get?
b He cooks 3 and finds they all have a single yolk. He argues that he now has a 1 in 3 chance of a double yolk from the remaining eggs. Explain why he is wrong.

★9 John has a bag containing 6 red, 5 blue and 4 green balls. One ball is picked from the bag at random. What is the probability that the ball is

a red or blue **b** not blue **c** pink **d** red or not blue?

HOMEWORK 16F

1 Two dice are thrown together. Draw a probability diagram to show the total score.

a What is the probability of a score that is
i 7 **ii** 5 or 8 **iii** bigger than 9 **iv** between 2 and 5
v odd **vi** a non-square number?

2 Two dice are thrown. Draw a probability diagram to show the outcomes as a pair of co-ordinates.

What is the probability that

a the score is a 'double'
b at least one of the dice shows 3
c the score on one dice is three times the score on the other dice
d at least one of the dice shows an odd number
e both dice show a 5
f at least one of the dice will show a 5
g exactly one dice shows a 5?

3 Two dice are thrown. The score on one dice is doubled and the score on the other dice is subtracted.
Complete the probability space diagram.
For the event described above, what is the probability of a score of

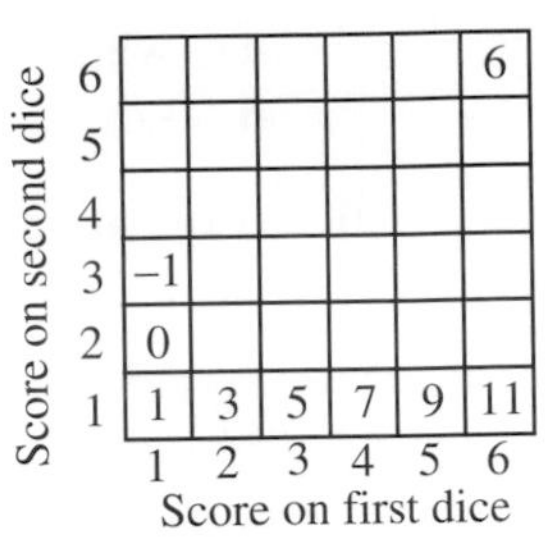

a 1
b a negative number
c an even number
d 0 or 1
e a prime number?

4 When two coins are tossed together, what is the probability of
a 2 heads or 2 tails **b** a head and a tail **c** at least 1 head?

5 When three coins are tossed together, what is the probability of
a 3 heads or 3 tails **b** 2 tails and 1 head **c** at least 1 head?

6 When a dice and a coin are thrown together, what is the probability of each of the following outcomes?
a You get a tail on the coin and a 3 on the dice.
b You get a head on the coin and an odd number on the dice.

★7 Max buys two bags of bulbs from his local garden centre. Each bag has 4 bulbs. Two bulbs are daffodils, one is a tulip and one is a hyacinth. Max takes one bulb from each bag.

Hyac				HH
Tulip	DT			
Daff				
Daff	DD	DD	TD	
	Daff	**Daff**	**Tulip**	**Hyac**

a There are six possible different pairs of bulbs. List them all.
b Complete the sample space diagram.
c What is the probability of getting two daffodil bulbs?
d Explain why the answer is not $\frac{1}{6}$.

HOMEWORK 16G

1 A dice is thrown twice. Copy and complete the tree diagram below to show all the outcomes.

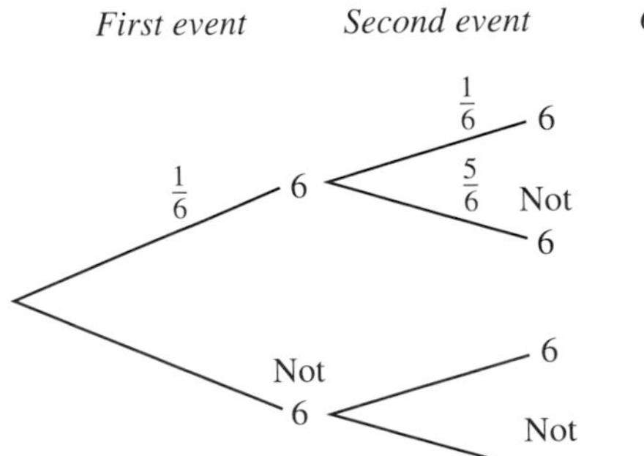

Use your tree diagram to work out the probability of
a getting two sixes **b** getting one six **c** getting no sixes.

2 A bag contains 3 red and 2 blue balls. A ball is taken out, replaced and then another ball is taken out.

a What is the probability that the first ball taken out will be red?

b Copy and complete the tree diagram below, showing the possible outcomes.

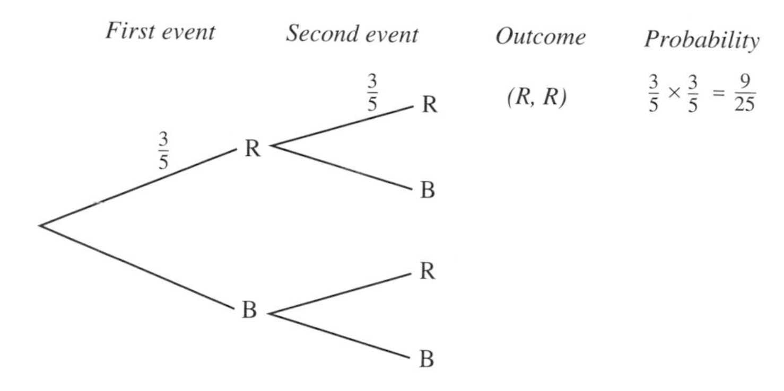

c Using the tree diagram, what is the probability of the following outcomes?

i 2 red balls. **ii** Exactly 1 red ball. **iii** I get at least one red ball.

3 A card is drawn from a pack of cards. It is replaced, the pack is shuffled and another card is drawn.

a What is the probability that either card was a Spade?

b What is the probability that either card was not a Spade?

c Draw a tree diagram to show the outcomes of two cards being drawn as described. Use the tree diagram to work out the probability that

i both cards will be Spades

ii at least one of the cards will be a Spade.

4 A bag of sweets contains 5 chocolates and 4 toffees.
I take 2 sweets out at random and eat them.

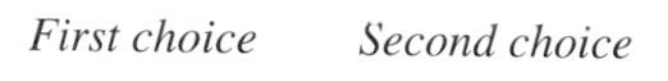

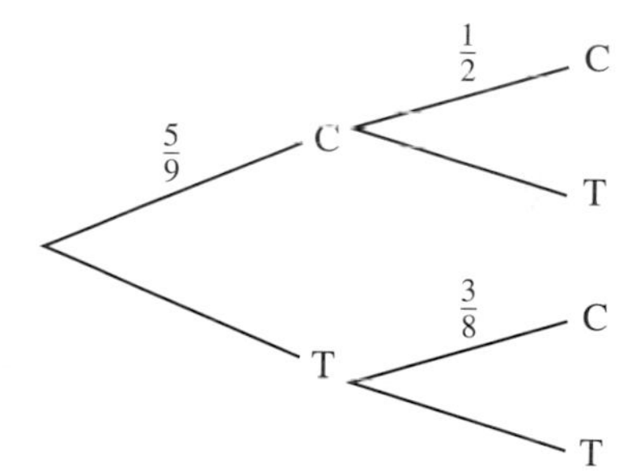

a What is the probability that the first sweet chosen is

i a chocolate **ii** a toffee?

b If the first sweet chosen is a chocolate,

i how many sweets are left to choose from

ii how many of them are chocolates?

c If the first sweet chosen is a toffee,

i how many sweets are left to choose from

ii how many of them are toffees?

d Copy and complete the tree diagram.

e Use the tree diagram to work out the probability that

i both sweets will be of the same type **ii** there is at least one chocolate chosen.

5 Thomas has to take a driving test which is in two parts. The first part is theoretical. He has a 0.4 chance of passing this. The second is practical. He has a 0.5 chance of passing this. Draw a tree diagram covering passing or failing the two parts of the test. What is the probability that he passes both parts?

6 Every Sunday morning Carol goes out for a run. She has 3 pairs of shorts of which 2 are red and 1 is blue. She has 5 T-shirts of which 3 are red and 2 are blue. Because she can't disturb her husband she gets dressed in the dark and picks a pair of shorts and a T shirt at random.

a What is the probability that the shorts are blue?

b Copy and complete this tree diagram.

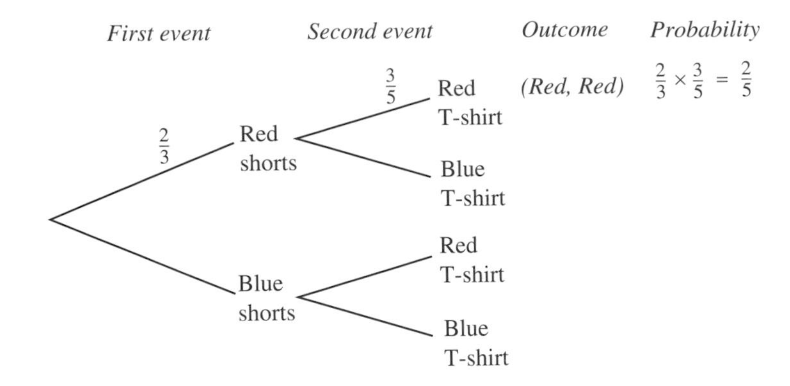

c What is the probability that Carol goes running in

i a matching pair of shorts and T shirt

ii a mis-match of shorts and T shirt

iii at least one red item?

★7 Bob has a bag containing 4 blue balls, 5 red balls and 1 green ball. Sally has a bag containing 2 blue balls and 3 red balls. The balls are identical except for the colour. Bob chooses a ball at random from his bag and Sally chooses a ball at random from her bag.

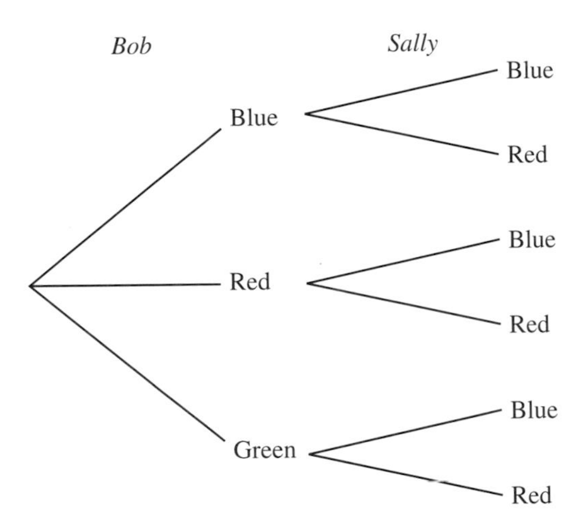

a On a copy of the tree diagram, write the probability of each of the events on the appropriate branch.

b Calculate the probability that both Bob and Sally will choose a blue ball.

c Calculate the probability that the ball chosen by Bob will be a different colour from the ball chosen by Sally.

Homework 16H

1 Ahmed throws a dice twice. The dice is biased so the probability of getting a six is $\frac{1}{4}$. What is the probability that he gets

a two sixes **b** exactly one six?

2 Betty draws a card from a pack of cards, replaces it, shuffles the pack and then draws another card. What is the probability that the cards are

a both Hearts **b** a Heart and a Spade (in any order)?

3 Colin draws a card from a pack of cards, does not replace it and then draws another card. What is the probability that the cards are

a both Hearts **b** a Heart and a Spade (in any order)?

4 I throw a dice three times. What is the probability of getting a total score of 17 or 18?

5 A bag contains 7 white beads and 3 black beads. I take out a bead at random, replace it and take out another bead. What is the probability that

a both beads are black **b** one bead is black and the other white (in any order)?

6 A bag contains 7 white beads and 3 black beads. I take out a bead at random, do not replace it and take out another bead. What is the probability that

a both beads are black **b** one bead is black and the other white (in any order)?

★7 When I answer the telephone the call is never for me. Half the calls are for my daughter Janette. One-third of them are for my son Glen. The rest are for my wife Barbara.

a I answer the telephone twice this evening. Calculate the probability that

i the first call will be for Barbara **ii** both calls will be for Barbara.

b The probability that both these calls are for Janette is $\frac{1}{4}$. The probability that they are both for Glen is $\frac{1}{9}$. Calculate the probability that either they are both for Janette or both for Glen.

HOMEWORK 16 I

1 Steve eats in the school canteen 5 days a week. The probability that there is turnip on the menu on any day is $\frac{1}{4}$.

a What is the probability that in 5 days there is no turnip on the menu?

b What is the probability that Steve gets turnip at least once in 5 days?

2 Three coins are thrown. What is the probability of

a 3 tails **b** at least 1 head?

3 Adam is a talented all-round athlete. He is entered for the 100 m race, the javelin and the high jump. The probability that he wins each of these events is $\frac{4}{5}$, $\frac{3}{4}$ and $\frac{1}{2}$ respectively.

a What is the probability that he doesn't win any event?

b What is the probability that he wins at least one event?

4 A bag contains 7 red and 3 blue balls. A ball is taken out and replaced. Another ball is taken out. What is the probability that

a both balls are red **b** both balls are blue **c** at least one is red?

5 A bag contains 7 red and 3 blue balls. A ball is taken out and not replaced. Another ball is taken out. What is the probability that

a both balls are red **b** both balls are blue **c** at least one is red?

6 **a** A coin is thrown three times. What is the probability of

i 3 heads **ii** no heads **iii** at least one head?

b A coin is thrown four times. What is the probability of

i 4 heads **ii** no heads **iii** at least one head?

c A coin is thrown five times. What is the probability of

i 5 heads **ii** no heads **iii** at least one head?

d A coin is thrown n times. What is the probability of
i n heads ii no heads iii at least one head?

7 A class has 12 boys and 14 girls. Three of the class are to be chosen at random to do a task for their teacher. What is the probability that
a all three are girls b at least one is a boy?

★8 From three boys and two girls, two children are to be chosen to present a bouquet of flowers to the Mayoress. The children are to be selected by drawing their names out of a hat. What is the probability that
a there will be a boy and a girl chosen
b there will be two boys chosen
c there will be at least one girl chosen?

HOMEWORK 16J

1 A bag contains 2 black balls and 5 red balls. A ball is taken out and replaced. This is repeated twice.What is the probability that
a all 3 are black b exactly 2 are black
c exactly 1 is black d none is black?

2 A bag contains 2 blue balls and 5 white balls. A ball is taken out but not replaced. This is repeated twice. What is the probability that
a all 3 are blue b exactly 2 are blue
c exactly 1 is blue d none is blue?

3 A dice is thrown 3 times. What is the probability that
a 4 sixes are thrown b no sixes are thrown c exactly one six is thrown?

4 Ann is late for school with a probability of 0.7. Bob is late with a probability of 0.6. Cedrick is late with a probability of 0.3. On any particular day what is the probability of
a exactly one of them being late b exactly two of them being late?

5 Daisy takes three AS exams in Mathematics. The probability that she will pass Pure 1 is 0.9. The probability that she will pass Statistics 1 is 0.65. The probability she will pass Discrete 1 is 0.95. What is the probability that she will pass
a all three modules b exactly two modules c at least two modules?

6 6 out of 10 cars in Britain run on petrol. The rest use diesel fuel. Three cars can be seen approaching in the distance.
a What is the probability that the first one uses petrol?
b What is the probability that exactly two of them use diesel fuel?
c Explain why, if the first car uses petrol, the probability of the second car using petrol is still 6 out of 10.

★7 In a class of 30 pupils, 21 have dark hair, 7 have fair hair and 2 have red hair. Two pupils are chosen at random to collect homework. What is the probabilty that they
a both have fair hair
b each have hair of a different colour?
c If 3 pupils are chosen, what is the probability that exactly 2 have dark hair?

HOMEWORK 16K

1 A box contains 8 red and 7 yellow balls. One is taken out and not replaced. Another is taken out.

a If the first ball taken out is red, what is the probability that the second ball is
i red **ii** yellow?

b If the first ball taken out is yellow, what is the probability that the second ball is
i red **ii** yellow?

2 Ann has a bargain box of tins. They are unlabelled but she knows that 8 tins contain salmon and 4 contain pears.

a She opens 2 tins. What is the probability that
i they are both salmon **ii** they are both pears?

b What is the probability that she has to open 2 tins before she gets a tin of pears?

c What is the probability that she has to open 3 tins before she gets a tin of pears?

d What is the probability that she will get a tin of salmon if she opens 5 tins?

3 A bag contains 6 black balls and 4 red balls. A ball is taken out and not replaced. This is repeated twice. What is the probability that

a all 3 are black **b** exactly 2 are black
c exactly 1 is black **d** none is black?

4 On my way to work, I pass two sets of traffic lights. The probability that the first is green is $\frac{3}{4}$. If the first is green, the probability that the second is green is $\frac{1}{2}$. If the first is red, the probability that the second is green is $\frac{1}{3}$. What is the probability that

a both are green **b** none is green
c exactly one is green **d** at least one is green?

5 A hand of five cards is dealt. What is the probability that

a all five are Hearts
b all five are the same suit
c they are four Kings and any other card
d they are four of a kind and any other card?

★6 Mrs White and Mr Black are solicitors. During office hours Mrs White is out of the office $\frac{1}{2}$ of the time, while, independently, Mr Black is out of the office $\frac{5}{8}$ of the time.

a Calculate the probability that either Mrs White or Mr Black, but not both, is in the office.

b The office is open from 9 a.m. to 5 p.m. Monday to Friday. For how many of these hours each week, on average, are neither Mrs White nor Mr Black in the office?

HOMEWORK 16L

1 A bag contains 5 black and 3 white balls. Three balls are taken out one at a time.

a If the balls are put back each time, what is the probability of getting
i 3 black balls **ii** at least 1 black ball?

b If the balls are not put back each time, what is the probability of getting
i 3 black balls **ii** at least 1 black ball?

2 Two cards are drawn one at a time from a pack of cards. The cards are replaced each time. What is the probability that at least one of them is a Heart?

3 Two cards are drawn from a pack of cards. The cards are not replaced each time. What is the probability that at least one of them is a Heart?

4 A box contains 50 batteries. It is known that 20 of them are dead. John needs two batteries for his calculator. He takes three out. What is the probability that

a all three of them are dead **b** at least one of them works?

5 From the same box of 50 batteries (Question **4**), Janet takes out three batteries for her radio. The radio will work if two or three of the batteries are good. What is the probability that the radio will work?

6 Two tetrahedral (four-sided) dice, whose faces are numbered 1, 2, 3 and 4, are thrown together. The score on the dice is the uppermost face. Show by means of a sample space diagram that there are 16 outcomes with total scores from 2 to 8. What is the probability of

a a total score of 8 **b** at least one 3 on either dice?

7 Based on previous results the probability that Manchester United win is $\frac{2}{3}$, the probability they draw is $\frac{1}{4}$ and the probability they lose is $\frac{1}{12}$.
They play three matches. What is the probability that

a they win all three **b** they win exactly 2 matches

c they win at least one match?

8 In the Spark'n, Spit'n and Fizz'n machine, the probability that the Spark fails is 0.02. The probability that the Spit fails is 0.08 and the probability that the Fizz fails is 0.05.

a What is the probability that nothing fails?

b The machine will still work with one component out of action. What is the probability that it works?

9 On average, Steve is late for school two days each week (of 5 days).

a What is the probability that he is late on any one day?

b In a week of 5 days, what is the probability that

i he is late every day **ii** he is late exactly once

iii he is never late **iv** he is late at least once?

10 Six T-shirts are hung out at random on a washing line. Three are red and three are blue. Using R and B, write down all 20 possible combinations: for example, RRRBBB, RRBBBR, …. What is the probability of

a 3 red shirts being next to each other

b 3 blue shirts being next to each other?

11 Dan has 6 socks in a drawer, of which 4 are blue and 2 are black. He takes out 2 socks. What is the probability that

a both socks are blue **b** both socks are black

c he gets a pair of socks **d** at least one of the socks is blue?

12 One in nine people are left handed. Five people are in a room. What is the probability that

a all 5 are left-handed **b** all 5 are right-handed

c at least one of them is left-handed?

13 A maze is tilted slightly to the left. A ball is dropped into a maze at A. It can go left with a probability of $\frac{2}{3}$ and right with a probability of $\frac{1}{3}$.

a Show that the probability that the ball lands in slot 1 is $\frac{8}{27}$.

b Show that the probability that the ball lands in slot 2 is $\frac{12}{27}$.

c 54 balls are dropped into the maze at A. How many would you expect to land in slot 4?

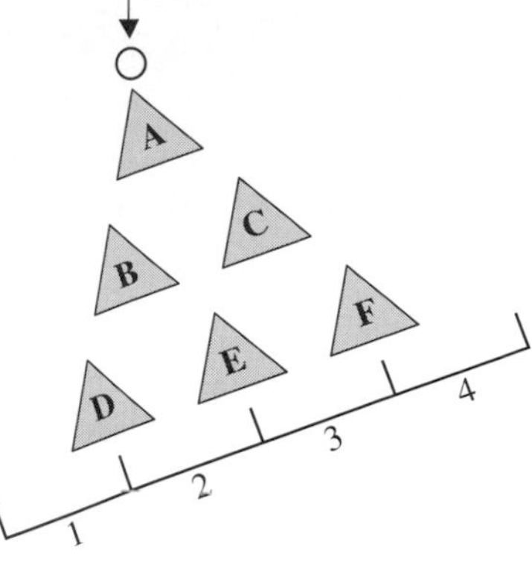

★**14** There are 5 yellow counters in a bag of 9 coloured counters. Three counters are taken out of the bag at random, one at a time and without replacement.

a Find the probability that exactly two of the three counters are yellow.

A group of students do an experiment with the bag of 9 coloured counters. They take two counters from the bag at random, one at a time and without replacement. They record the colour of each counter and then put the two counters back in the bag. They repeat this experiment 1000 times. They find that the relative frequency of taking two red counters from the bag is 0.083.

b **i** How many times were two red counters taken out of the bag?

ii Show that 3 is a good estimate of the most likely number of red counters in the bag.

Chapter 17 Algebra 3

Homework 17A

1 Use each of the following rules to write down the first five terms of a sequence.

a $3n + 1$ for $n = 1, 2, 3, 4, 5$ **b** $2n - 1$ for $n = 1, 2, 3, 4, 5$

c $4n + 2$ for $n = 1, 2, 3, 4, 5$ **d** $2n^2$ for $n = 1, 2, 3, 4, 5$

e $n^2 - 1$ for $n = 1, 2, 3, 4, 5$

2 Write down the first five terms of the sequence which has its nth term as

a $n + 2$ **b** $4n - 1$ **c** $4n - 3$ **d** $n^2 + 1$ **e** $2n^2 + 1$

3 **a** Write down the first six terms of the sequence of fractions $\dfrac{2n + 1}{n + 1}$ for $n = 1, 2, 3, 4, \ldots$

b By letting $n = 1000$ use your calculator to work out the value of the fraction as a decimal when $n = 1000$.

c What fraction do you think the sequence is heading towards?

★**4** **a** A sequence is given by $\frac{1}{2}, \frac{2}{3}, \frac{3}{4}, \frac{4}{5}, \ldots$.

Write down

i the 11th term of this sequence **ii** the nth term of this sequence.

b Each term of a second sequence is the reciprocal of the corresponding term of the sequence given in part **a**. Write down the first four terms of the second sequence.

HOMEWORK 17B

1 Find the nth term in each of these linear sequences.

a 5, 7, 9, 11, 13 … **b** 3, 7, 11, 15, 19, … **c** 6, 11, 16, 21, 26, …
d 3, 9, 15, 21, 27, … **e** 4, 7, 10, 13, 16, … **f** 3, 10, 17, 24, 31, …

2 Find the 50th term in each of these linear sequences.

a 3, 5, 7, 9, 11, … **b** 5, 9, 13, 17, 21, … **c** 8, 13, 18, 23, 28, …
d 2, 8, 14, 20, 26, … **e** 5, 8, 11, 14, 17, … **f** 2, 9, 16, 23, 30, …

3 For each sequence **a** to **f**, find

i the nth term **ii** the 100th term **iii** the term closest to 100

a 4, 7, 10, 13, 16, … **b** 7, 9, 11, 13, 15, … **c** 3, 8, 13, 18, 23, …
d 1, 5, 9, 13, 17, … **e** 2, 10, 18, 26, … **f** 5, 6, 7, 8, 9, …

4 A sequence of fractions is $\frac{3}{5}, \frac{5}{8}, \frac{7}{11}, \frac{9}{14}, \ldots$

a Find the nth term in the sequence. **b** Change each fraction to a decimal.
c What, as a decimal, will be the value of the **i** 100th term **ii** 1000th term?
d Use your answers to part **c** to predict what the 10 000th term and the millionth term are. (Check these out on your calculator.)

★5 a A number pattern begins 1, 1, 2, 3, 5, 8, …

i What is the next number in this pattern?
ii The number pattern is continued. Explain how you would find the tenth number in the pattern.

b Another number pattern begins 1, 5, 9, 13, 17, … . Write down, in terms of n, the nth term in this pattern.

HOMEWORK 17C

1 A pattern of shapes is built up from matchsticks as shown.

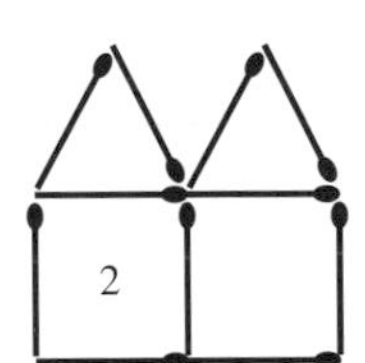

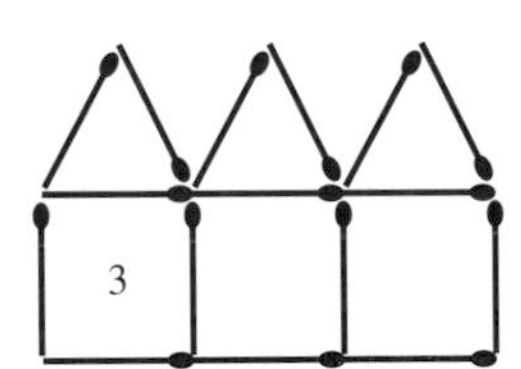

a Draw the 4th diagram.
b How many matchsticks are in the nth diagram?
c How many matchsticks are in the 25th diagram?
d With 200 matchsticks, which is the biggest diagram that could be made?

2 A pattern of hexagons is built up from matchsticks.

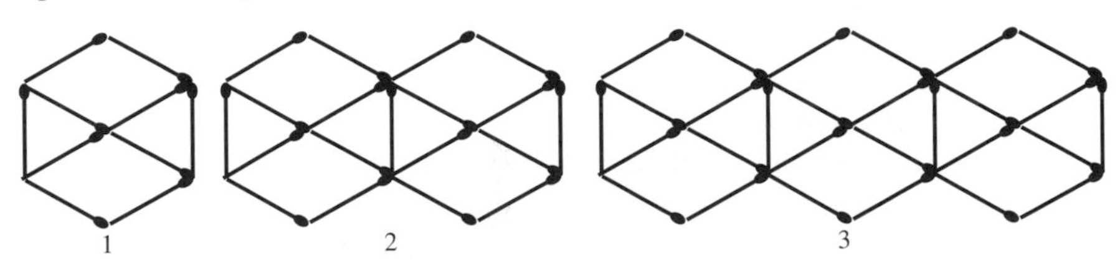

a Draw the 4th set of hexagons in this pattern.
b How many matchsticks are needed for the nth set of hexagons?

c How many matchsticks are needed to make the 60th set of hexagons?

d If there are only 100 matchsticks, which is the largest set of hexagons that could be made?

3 A conference centre had tables each of which could sit 3 people. When put together, the tables could seat people as shown.

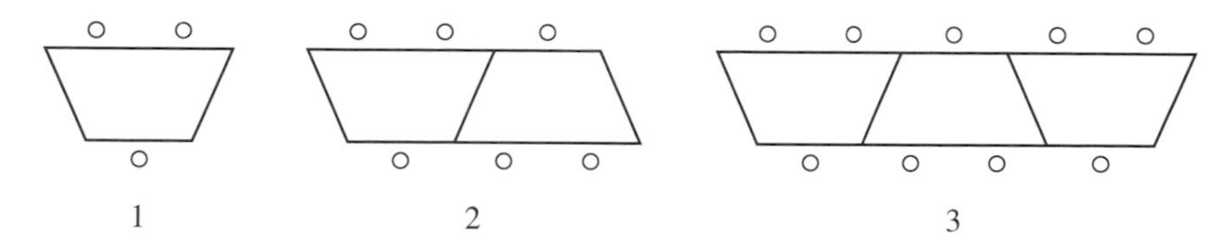

a How many people could be seated at 4 tables?

b How many people could be seated at n tables put together in this way?

c A conference had 50 people who wished to use the tables in this way. How many tables would they need?

4 A pattern of squares is put together as shown.

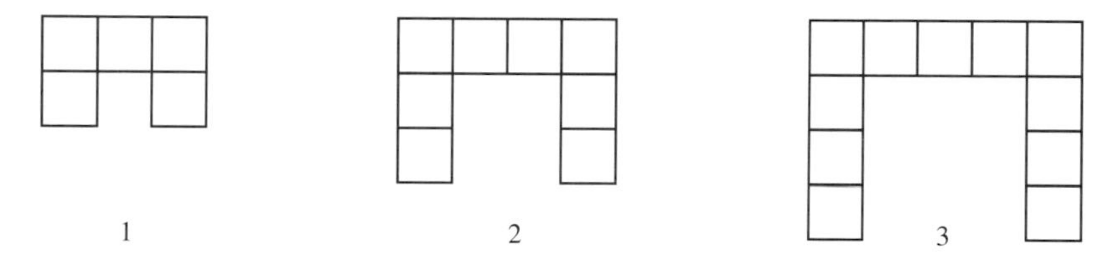

a Draw the 4th diagram.

b How many squares are in the nth diagram?

c How many squares are in the 50th diagram?

d With 300 squares, work out the number of the biggest diagram that could be made.

★5 Sheep enclosures are built using fences and posts.

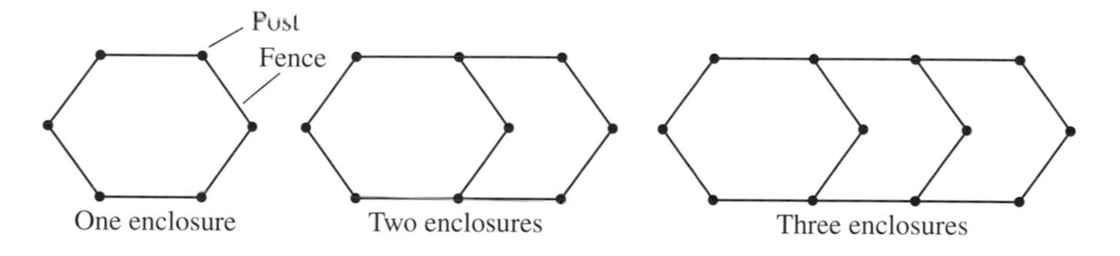

a **i** Sketch four enclosures in a row. **ii** Sketch five enclosures in a row.

b Copy and complete the table below.

Number of enclosures	1	2	3	4	5	6	7	8
Number of posts	6	9	12					

c Work out the number of posts needed for 20 enclosures in a row.

d Write down an expression to find the number of posts needed for n enclosures in a row.

HOMEWORK 17D

1 For each of the sequences **a** to **e**

i write down the next two terms **ii** find the nth term.

a 1, 4, 9, 16, 25, … **b** 2, 5, 10, 17, 26, … **c** 5, 8, 13, 20, 29, …

d 2, 8, 18, 32, 50, … **e** 21, 24, 29, 36, 45, …

2 For each of the sequences **a** to **e**

i write down the next two terms **ii** find the nth term.

a 4, 10, 18, 28, … **b** 15, 24, 35, 48, … **c** 2, 6, 12, 20, …

d 1, 3, 6, 10, … **e** 6, 12, 20, 30, …

3 Look at each of the following sequences to see whether the rule is linear, quadratic on n^2 alone or fully quadratic. Then

i write down the nth term **ii** write down the 50th term.

a 8, 13, 20, 29, … **b** 8, 11, 14, 17, … **c** 8, 15, 24, 35, …

d 0, 5, 12, 21, 32, 45, …

★4 The picture shows a pattern of cards.

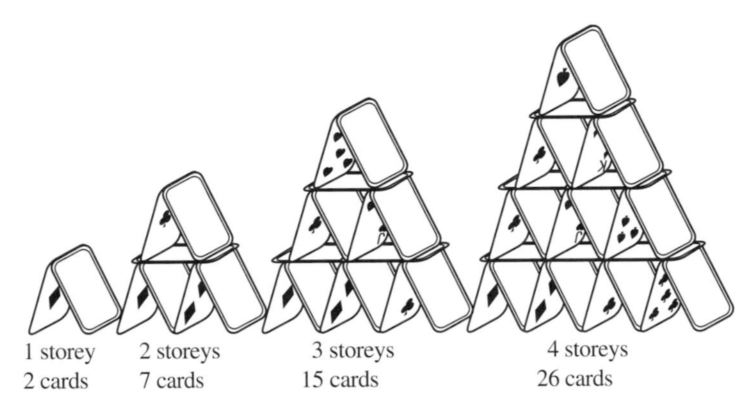

a The four-storey house of cards is to be made into a five-storey house of cards. How many more cards are needed?

b Look at the sequence 2, 7, 15, 26, …

i Calculate the sixth term in this sequence.

ii Explain how you found your answer.

HOMEWORK 17E

In questions 1 to 5, make the stated term the subject of each formula.

1 $4(x - 2y) = 3(2x - y)$ (x)

2 $p(a - b) = q(a + b)$ (a)

3 $A = 2ab^2 + ac$ (a)

4 $s(t + 1) = 2r + 3$ (r)

5 $st - r = 2r - 3t$ (t)

6 Make x the subject of these equations

a $ax = b - cx$ **b** $x(a - b) = x + b$

c $a - bx = dx - a$ **d** $x(c - d) = c(d - x)$

7 **a** The perimeter of the shape on the right is given by $P = 2\pi r + 4r$
Make r the subject of the formula.

b The area of the same shape is given by $A = \pi r^2 + 4r^2$
Make r the subject of this formula.

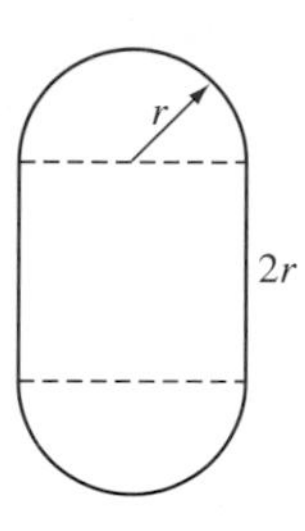

8 **a** Make x the subject of $y = \dfrac{x + 2}{x + 3}$

b Make x the subject of $y = \dfrac{2 - 3x}{x - 1}$

★9 Make a the subject of $a = \dfrac{2 + 3a}{b - 2}$

10 The resistance when two resistors with values a and b are connected in parallel is given by

$$R = \frac{ab}{a + b}$$

a Make b the subject of the formula.

b Write down the formula when a is the subject.

Chapter 18 Dimensional analysis

Homework 18A

Find an expression for the perimeter of each of these shapes.

1

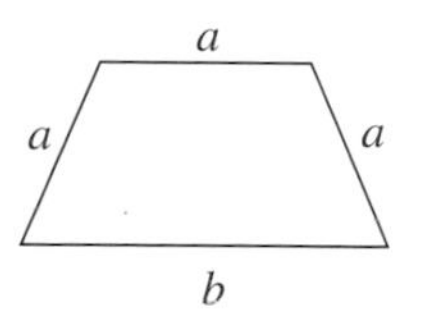

2

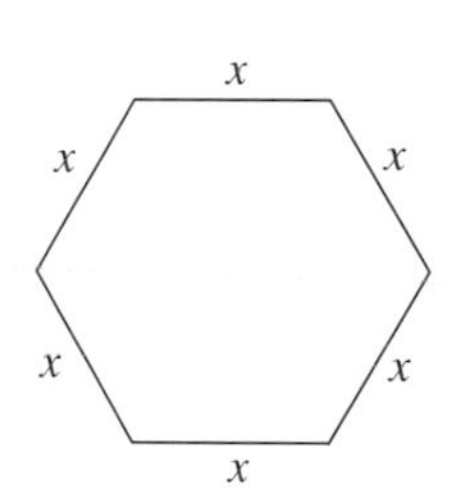

3

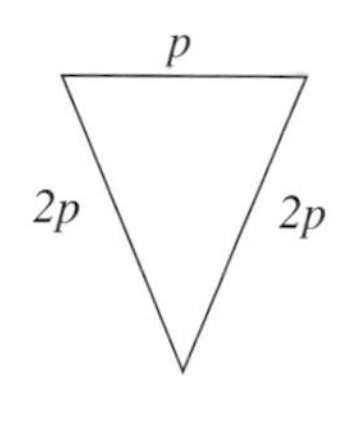

4

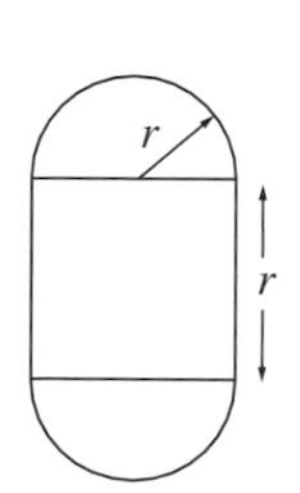

5

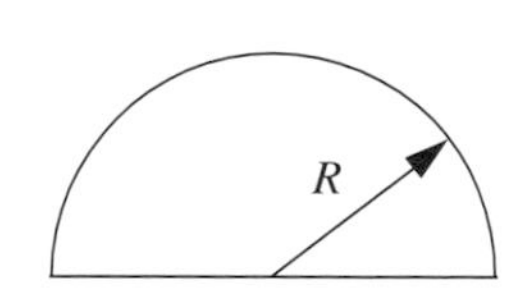

6

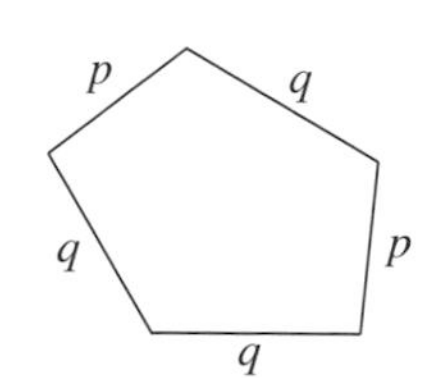

★7 The diagram shows a child's play brick in the shape of a prism.
The following formulae represent certain quantities connected with this prism.

$\pi ab \quad \pi(a + b) \quad \pi abl \quad \pi(a + b)l$

Which of the formulae represents a length?

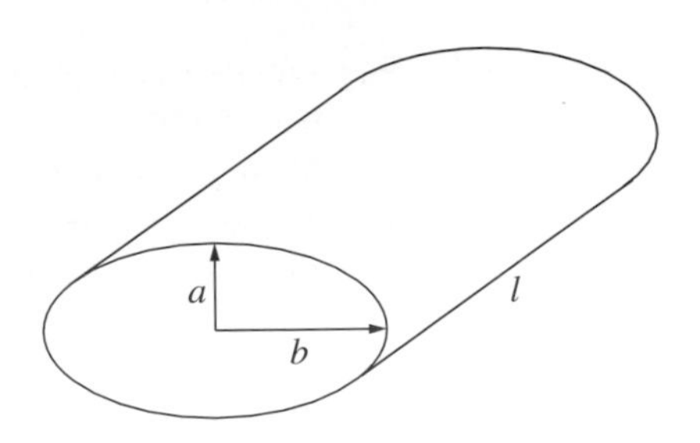

HOMEWORK 18B

Find a formula for each of these areas.

1

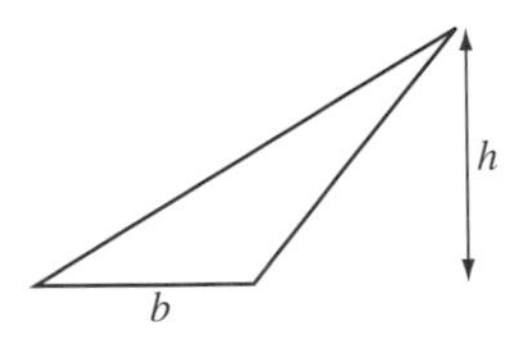

2

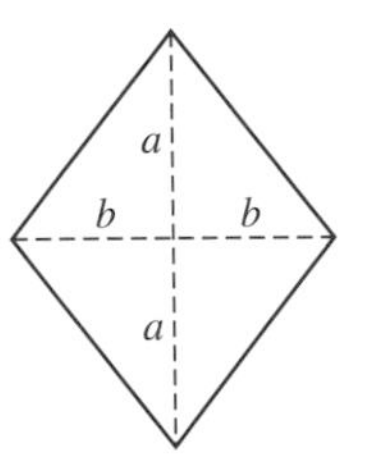

3

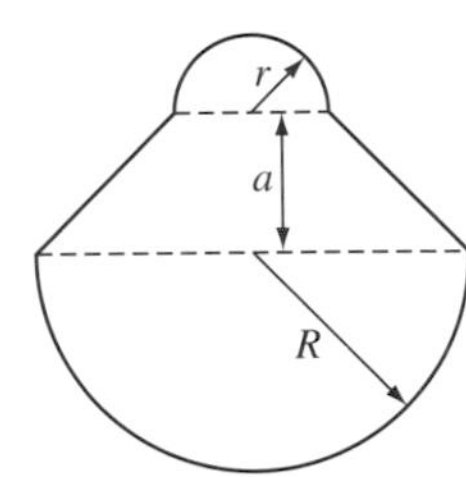

4

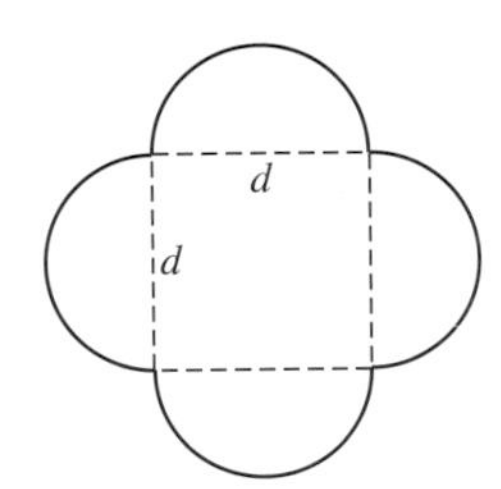

5

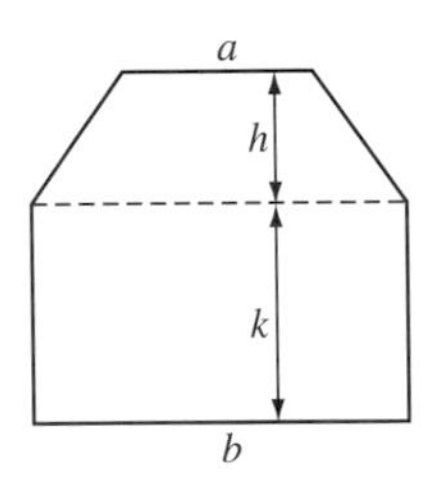

6

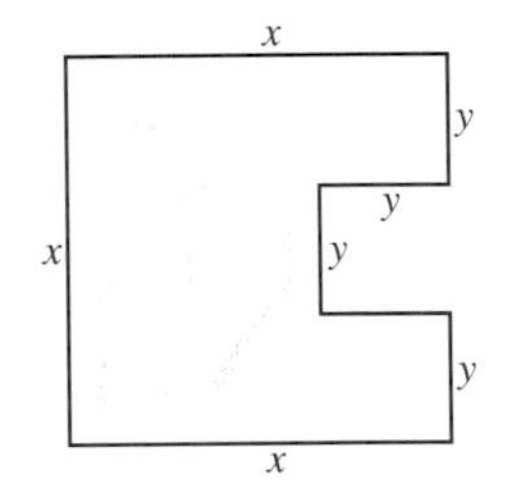

★7 One of the formulae in the list below can be used to calculate the area of material in a lampshade. Which formula is it?

a $\pi h(a + b)^2$ **b** $\pi h^2(a + b)$ **c** $\pi h(a + b)$ **d** $\pi h^2(a + b)^2$

HOMEWORK 18C

Find a formula for each of these volumes.

1

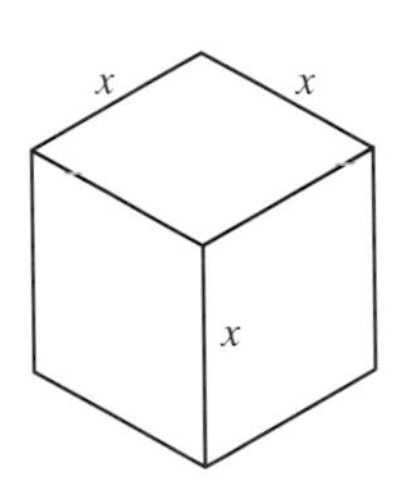

2

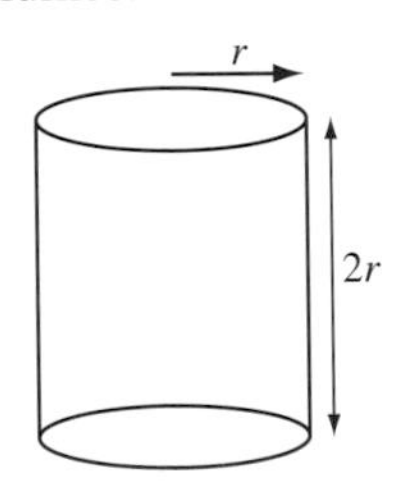

3

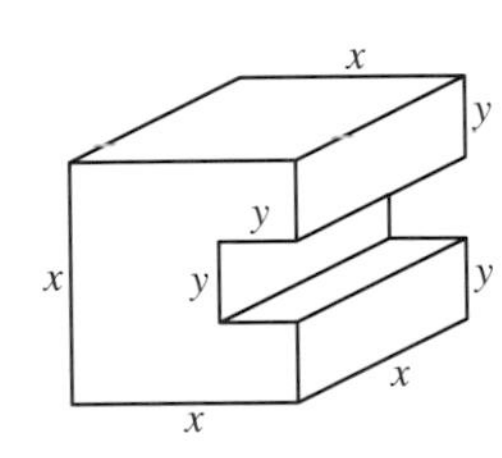

4

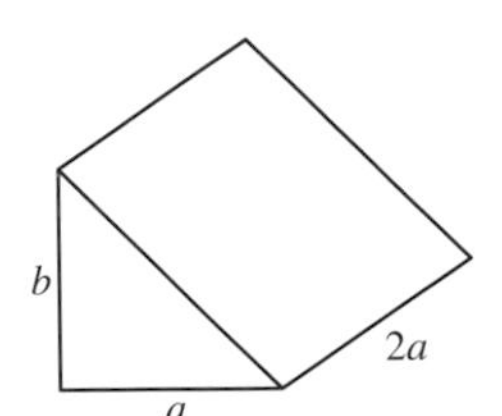

5

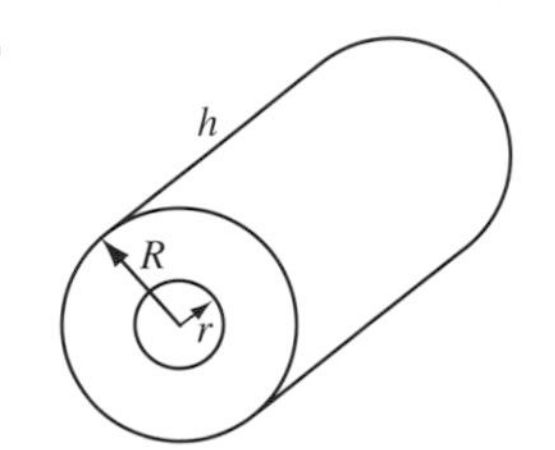

6

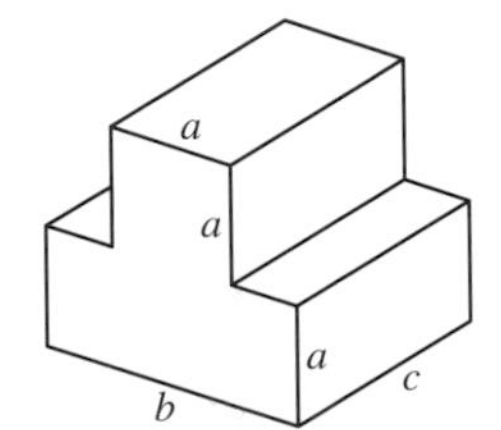

★**7** The dimensions of four cuboids are shown.

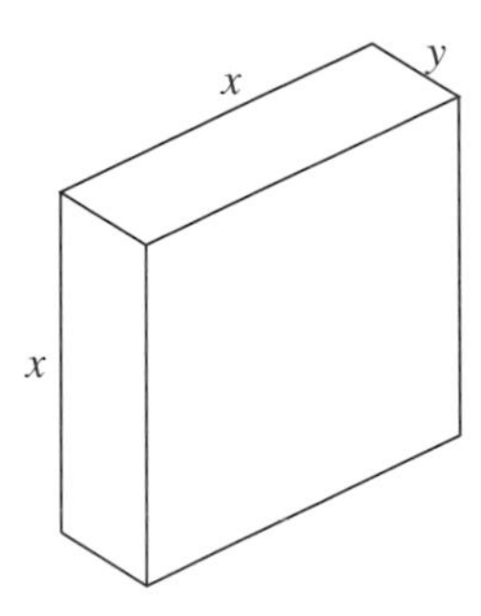

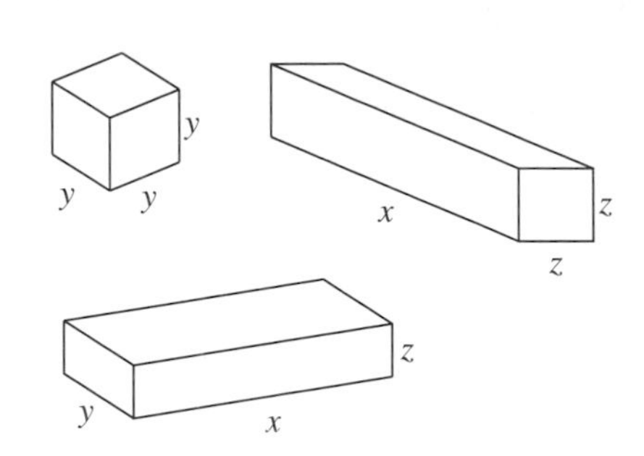

The expressions x^2y, xz and $2(x + y)$ give the perimeter of a face of one of the cuboids, or the area of a face of one of the cuboids, or the volume of one of the cuboids. Complete the statements below with the word perimeter, area or volume.

a x^2y gives a …… **b** $2(x + y)$ gives a …… **c** xz gives an …….

HOMEWORK 18D

1 Indicate by L, A, V or [L], [L^2], [L^3] whether the following quantities are lengths, areas or volumes, or none of these (N).

a 1 yard
b 12 hectares
c Paint in a can
d Weight of this book
e Thickness of this book
f Amount of paper used to print this book
g 110 decibels
h Amount of water flowing from a tap
i Latitude and longitude
j m s^{-2}

2 Each of these represents a length, an area or a volume. Indicate by writing L, A or V which it is.

a $ab + cd$ **b** $a^2b + c^2d$ **c** $ab(c + d)$ **d** $\frac{4}{3}\pi r^3$

e $\pi r^2 + 2\pi rh$ **f** $\frac{2}{3}\pi r^3 + \pi r^2 h$ **g** $\dfrac{ab + cb}{b}$ **h** $\dfrac{a^3b + b^3a}{c}$

★**3** This diagram shows the cross-section of a barn. A farmer wants to estimate the area of a cross-section of a warehouse.

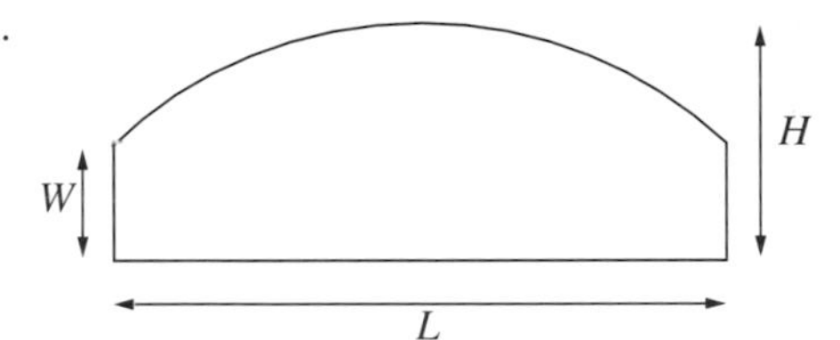

a Explain why the formula

$$A = \frac{H + 2W + L}{5}$$

cannot be a suitable formula for him to use.

b Look at these formulae.

i $A = \dfrac{4WHL}{5}$ **ii** $A = \dfrac{L(4H + 2W)}{5}$ **iii** $A = \dfrac{LW + 4H}{5}$

One of these formulae can be used to estimate the area of the cross-section. Which is it? Give a reason for your answer.

Homework 18E

1 Indicate whether each of these formulae is consistent (C) or inconsistent (I).

a $a^2 + cd$ **b** $a^2b + cb$ **c** $ab(c^2 + d)$ **d** $r^2 + \frac{4}{3}\pi r$

e $\frac{2}{3}\pi r^3 + 2\pi rh$ **f** $\pi(a^3 + b^3)$ **g** $(a^2 + b)(a + b^2)$ **h** $(x + y)(x - y)$

2 **i** Write down whether each formula is consistent (C) or inconsistent (I).

ii When it is consistent, say whether it represents a length (L), an area (A) or a volume (V).

a $\pi a^2 + cd$ **b** $\dfrac{\pi(ab + cd)}{a}$ **c** $\pi r^2h + 2\pi r$ **d** $(\pi + 2)a$

e $abc + bca + cab$ **f** $\dfrac{\pi ab + \pi r^3}{a}$ **g** $a(b + c) + \pi r^2$ **h** $\dfrac{\pi(a^2 + b^2)}{c^2}$

3 What power * would make each formula consistent?

a $\pi r^* h + \frac{4}{3}\pi r^3$ **b** $\dfrac{a^* b + r^2}{c}$ **c** $\pi a^2(b^* + ab)$ **d** $2a^* b + \pi r^3$

★4 In the following formulae r and h represent a length. For each formula, state whether it represents a length, an area, a volume or none of them.

a $2\pi r^2h$ **b** $\pi r^2(r + h)$ **c** $\sqrt{r^2 + h^2}$ **d** $\dfrac{r^2}{2\pi h}$ **e** $3\pi + r$

Chapter 19 Variation

Homework 19A

In each case, first find k, the constant of proportionality, and then the formula connecting the variables.

1 T is directly proportional to M. If $T = 30$ when $M = 5$, find

a T when $M = 4$ **b** M when $T = 75$

2 W is directly proportional to F. If $W = 54$ when $F = 3$, find

a W when $F = 4$ **b** F when $W = 90$

3 P is directly proportional to A. If $P = 50$ when $A = 2$, find

a P when $A = 5$ **b** A when $P = 150$

4 A is directly proportional to t. If $A = 45$ when $t = 5$, find

a A when $t = 8$ **b** t when $A = 18$

5 Q varies directly with P. If $Q = 200$ when $P = 5$, find

a Q when $P = 3$ **b** P when $Q = 300$

6 The distance covered by a car is directly proportional to the time taken. The train travels 135 miles in 3 hours.

a What distance will the train cover in 4 hours?

b What time will it take for the train to cover 315 miles?

7 The cost of petrol is directly proportional to the amount put in the tank. When 40 litres is used, it costs £32.00. How much

a will it cost for 30 litres **b** would be delivered if the cost were £38.40?

8 The number of people who can meet safely in a room is directly proportional to the area of the room. A room with an area of $200\,m^2$ is safe for 50 people.

a How many people can safely meet in a room of area $152\,m^2$?

b A committee has 24 members. What is the smallest room area in which they could safely meet?

★9 A man lays 36 paving stones in 3 hours. Working at the same rate how long would he take to lay 45 paving stones.

HOMEWORK 19B

In each case, first find k, the constant of proportionality, and then the formula connecting the variables.

1 T is directly proportional to x^2. If $T = 40$ when $x = 2$, find

a T when $x = 5$ **b** x when $T = 400$

2 W is directly proportional to M^2. If $W = 10$ when $M = 5$, find

a W when $M = 4$ **b** M when $W = 64$

3 A is directly proportional to r^2. If $A = 96$ when $r = 4$, find

a A when $r = 5$ **b** r when $A = 12$

4 E varies directly with $\sqrt{C}$. If $E = 60$ when $C = 36$, find

a E when $C = 49$ **b** C when $E = 160$

5 X is directly proportional to $\sqrt{Y}$. If $X = 80$ when $Y = 16$, find

a X when $Y = 100$ **b** Y when $X = 48$

6 H is directly proportional to $\sqrt{t}$. If $H = 20.8$ when $t = 4$, find

a H when $t = 1.5$ **b** t when $H = 77.4$

7 P is directly proportional to f^3. If $P = 500$ when $f = 5$, find

a P when $f = 4$ **b** f when $P = 50$

8 The temperature, in °C, in an experiment varied directly with the square of the pressure, in atmospheres. The temperature was 30 °C when the pressure was 6 atm. What will

a the temperature be at 3 atm **b** the pressure be at 90 °C?

9 The weight, in grams, of marbles varies directly with the cube of the radius. A ball bearing of radius 5 mm has a weight of 10 g.

a What will a ball bearing of radius 10 mm weigh?

b A ball bearing has a weight of 5.12 g. What is its radius?

10 The energy, in joules, of a particle varies directly with the square of its speed. A particle moving at 10 m/s has an energy of 500 joules.

a How much energy has a particle moving at 5 m/s?

b How fast is a particle moving if it has 1000 joules of energy?

★11 The distance that can be seen out to sea on a clear day varies directly as the square root of the height above sea level. At a height of 6 metres above sea level, you can see 10 km on a clear day. What distance will you be able to see on a clear day from a height of 60 metres above sea level?

HOMEWORK 19C

In each case, first find the formula connecting the variables.

1 T is inversely proportional to m. If $T = 7$ when $m = 4$, find
 a T when $m = 5$ **b** m when $T = 56$

2 W is inversely proportional to x. If $W = 6$ when $x = 15$, find
 a W when $x = 3$ **b** x when $W = 10$

3 M varies inversely with t^2. If $M = 10$ when $t = 2$, find
 a M when $t = 4$ **b** t when $M = 160$

4 C is inversely proportional to f^2. If $C = 20$ when $f = 3$, find
 a C when $f = 5$ **b** f when $C = 720$

5 W is inversely proportional to $\sqrt{T}$. If $W = 8$ when $T = 36$, find
 a W when $T = 25$ **b** T when $W = 0.75$

6 H varies inversely with $\sqrt{g}$. If $H = 20$ when $g = 16$, find
 a H when $g = 1.25$ **b** g when $H = 40$

7 The brightness of a bulb decreases inversely with the square of the distance away from the bulb. The brightness is 5 candle power at a distance of 10 metres. What is the brightness at a distance of 5 metres?

8 The density of a series of spheres with the same weight is inversely proportional to the cube of the radius. A sphere with a density of 10 g/cm³ has a radius of 5 cm.
 a What would be the density of a sphere with a radius of 10 cm?
 b If the density was 80 g/cm³ what would the radius of the sphere be?

★9 Given that y is inversely proportional to the square of x, and that y is 12 when $x = 4$,
 a find an expression for y in terms of x.
 b Calculate **i** y when $x = 6$ **ii** x when $y = 36$

Chapter 20 Number and limits of accuracy

HOMEWORK 20A

1 Write down the limits of accuracy for each of the following values which are to the given degree of accuracy.
 a 7 cm (1 sf) **b** 18 kg (2 sf) **c** 30 min (2 sf)
 d 747 km (3 sf) **e** 9.8 m (1 dp) **f** 32.1 kg (1 dp)
 g 3.0 h (1 dp) **h** 90 g (2 sf) **i** 4.20 mm (2 dp)
 j 2.00 kg (2 dp) **k** 34.57 min (2 dp) **l** 100 m (2 sf)

★2 Round off these numbers to the degree of accuracy given.
 a 45.678 to 3 sf **b** 19.96 to 2 sf **c** 0.3213 to 2 dp

HOMEWORK 20B

1 Write down the upper and lower bounds of each of these values given to the accuracy stated.

a	6 m (1 sf)	**b**	34 kg (2 sf)	**c**	56 min (2 sf)	**d**	80 g (2 sf)
e	3.70 m (2 dp)	**f**	0.9 kg (1 dp)	**g**	0.08 s (2 dp)	**h**	900 g (2 sf)
i	0.70 m (2 dp)	**j**	360 d (3 sf)	**k**	17 weeks (2 sf)	**l**	200 g (2 sf)

2 Billy has 20 identical bricks. Each brick is 15 cm long measured to the nearest centimetre.

a What is the greatest length of one brick?

b What is the smallest length of one brick?

c If the bricks are put end to end, what is the greatest possible length of all the bricks?

d If the bricks are put end to end, what is the least possible length of all the bricks?

★3 Cars are timed over a stretch of road using a digital timer which counts in tenths of a second. The digits change at the end of each complete tenth of a second. The time for one car is shown as 41.8 seconds.

a Complete the inequality below to show the possible times taken by the car.

...... ≤ time of car in seconds < ......

b The stretch of road is 800 m long, correct to the nearest 10 metres. Calculate the minimum speed the car could have been travelling. Give your result in metres per second, correct to 2 decimal places.

c Another car's speed is calculated as 31.73 metres per second (correct to 2 decimal places). Convert this speed to miles per hour. [1 metre per second = 2.237 miles per hour (correct to 3 decimal places).] Give your answer to an appropriate degree of accuracy. Show clearly why it is not possible to give a more exact answer.

HOMEWORK 20C

1 For each of these rectangles, find the limits of accuracy of the area. The accuracy of each measurement is given.

a 3 cm × 8 cm (nearest cm) **b** 3.2 cm × 6.4 cm (1 dp)

c 7.86 cm × 18.78 cm (2 dp)

2 A rectangular garden has sides of 8 m and 5 m, measured to the nearest metre.

a Write down the limits of accuracy for each length.

b What is the maximum area of the garden?

c What is the minimum perimeter of the garden?

3 A playground is measured as 32 m by 45 m, to the nearest metre. Calculate the limits of accuracy for the area of the playground.

4 The measurements, to the nearest centimetre, of a box are given as 12 cm × 8 cm × 5 cm. Calculate the limits of accuracy for the volume of the box.

5 The area of a field is given as 400 m^2, to the nearest 10 m^2. One length is given as 24 m, to the nearest metre. Find the limits of accuracy for the other length of the field.

6 In triangle ABC, AB = 8 cm, BC = 6 cm, and ∠ABC = 42°. All the measurements are given to the nearest unit.

a Calculate the limits of accuracy for the area of the triangle.

b Calculate the smallest possible length of AC.

7 A stop-watch records the time for the winner of a 100-metre race as 12.3 seconds, measured to the nearest one-tenth of a second.

a What are the greatest and least possible times for the winner?

b The length of the 100-metre track is correct to the nearest 1 centimetre. What are the greatest and least possible lengths of the track?

c What is the fastest possible average speed of the winner?

8 A cube has a volume of $27\,cm^3$, to the nearest cm^3. Find the range of possible values of the side length of the cube.

9 A cube has a volume of $125\,cm^3$, to the nearest $1\,cm^3$. Find the limits of accuracy of the area of one side of the square.

★10 In the triangle ABC, the length of side AB is 42 cm to the nearest centimetre. The length of side AC is 35 cm to the nearest centimetre. The angle C is 61° to the nearest degree. What is the largest possible size that angle B could be?

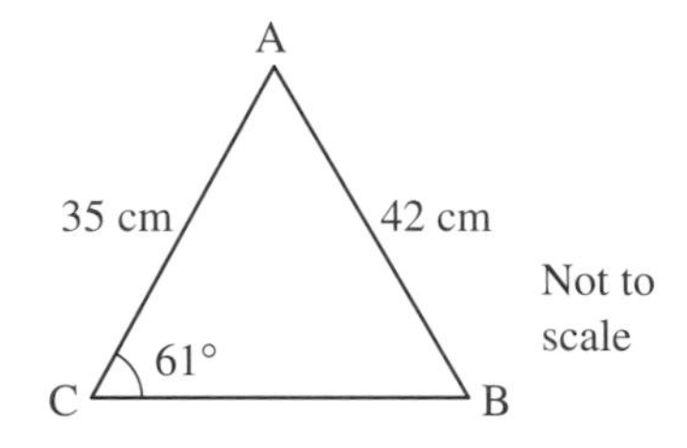

Chapter 21 Inequalities and regions

HOMEWORK 21A

1 Solve the following linear inequalities.

a	$x + 3 < 8$	**b**	$t - 2 > 6$	**c**	$p + 3 \geq 11$
d	$4x - 5 < 7$	**e**	$3y + 4 \leq 22$	**f**	$2t - 5 > 13$
g	$\frac{x+3}{2} < 8$	**h**	$\frac{y+4}{3} \leq 5$	**i**	$\frac{t-2}{5} \geq 7$
j	$2(x - 3) < 14$	**k**	$4(3x + 2) \leq 32$	**l**	$5(4t - 1) \geq 30$
m	$3x + 1 \geq 2x - 5$	**n**	$6t - 5 \leq 4t + 3$	**o**	$2y - 11 \leq y - 5$
p	$3x + 2 \geq x + 3$	**q**	$4w - 5 \leq 2w + 2$	**r**	$2(5x - 1) \leq 2x + 3$

2 Write down the values of x that satisfies each of the following.

a $x - 2 \leq 3$, where x is a positive integer.

b $x + 3 < 5$, where x is a positive, even integer.

c $2x - 14 < 38$, where x is a square number.

d $4x - 6 \leq 15$, where x is a positive, odd number.

e $2x + 3 < 25$, where x is a positive, prime number.

3 Solve the following linear inequalities.

a	$9 < 4x + 1 < 13$	**b**	$2 < 3x - 1 < 11$	**c**	$-3 < 4x + 5 \leq 21$
d	$2 \leq 3x - 4 < 15$	**e**	$10 \leq 2x + 3 < 18$	**f**	$-5 \leq 4x - 7 \leq 8$
g	$3 \leq 5x - 7 \leq 13$	**h**	$8 \leq 2x + 3 < 19$	**i**	$7 \leq 5x + 3 < 24$

HOMEWORK 21B

1 Write down the inequality that is represented by each diagram below.

a
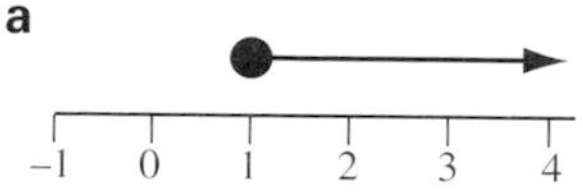

b
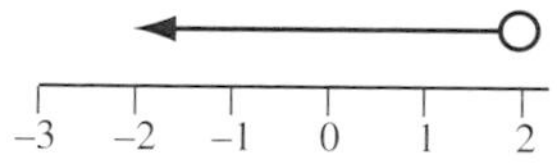

c
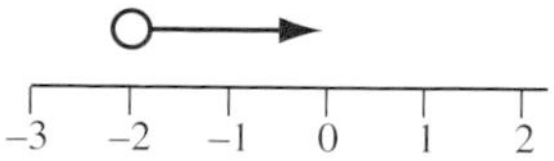

d
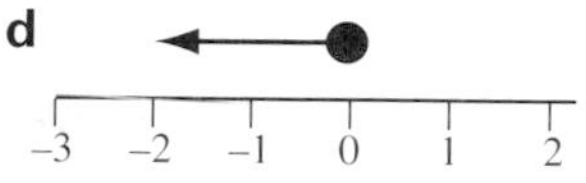

e
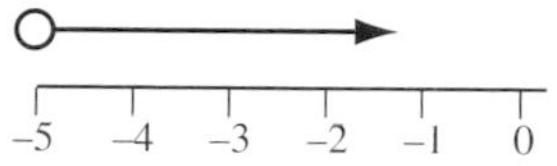

f
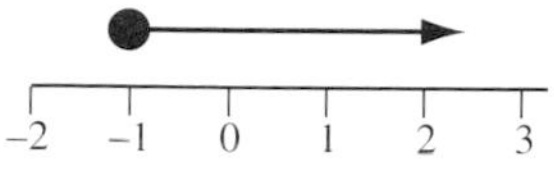

2 Draw diagrams to illustrate the following.

a $x \leq 2$ **b** $x > -3$ **c** $x \geq 0$ **d** $x < 4$

e $x \geq -3$ **f** $1 < x \leq 4$ **g** $-2 \leq x \leq 4$ **h** $-2 < x < 3$

3 Solve the following inequalities and illustrate their solutions on number lines.

a $x + 5 \geq 9$ **b** $x + 4 < 2$ **c** $x - 2 \leq 3$ **d** $x - 5 > -2$

e $4x + 3 \leq 9$ **f** $5x - 4 \geq 16$ **g** $2x - 1 > 13$ **h** $3x + 6 < 3$

i $3(2x + 1) < 15$ **j** $\frac{x+1}{2} \leq 2$ **k** $\frac{x-3}{3} > 7$ **l** $\frac{x+6}{4} \geq 1$

4 Solve the following inequalities and illustrate their solutions on number lines.

a $\frac{5x+2}{2} > 3$ **b** $\frac{3x-4}{5} \leq 1$ **c** $\frac{4x+3}{2} \geq 11$ **d** $\frac{2x-5}{4} < 2$

e $\frac{8x+2}{3} \leq 2$ **f** $\frac{7x+9}{5} > -1$ **g** $\frac{x-2}{3} \geq -3$ **h** $\frac{5x-2}{4} \leq -1$

HOMEWORK 21C

Solve the following inequalities, showing their solutions on number lines.

1 $x^2 \leq 9$ **2** $x^2 > 36$ **3** $x^2 < 100$ **4** $x^2 \geq 4$

5 $x^2 \geq 25$ **6** $x^2 - 1 > 15$ **7** $x^2 + 2 \leq 11$ **8** $x^2 + 3 > 3$

9 $x^2 + 6 > 7$ **10** $x^2 - 4 \geq 21$ **11** $2x^2 - 3 > 5$ **12** $x^2 - 3 < 6$

13 $5x^2 + 3 \leq 17$ **14** $2x^2 - 5 < 27$ **15** $3x^2 - 19 \geq 56$ **16** $x^2 \geq 144$

17 $x^2 < 0.16$ **18** $x^2 \geq 1.21$ **19** $x^2 - 5 \leq 88$ **20** $x^2 > 0.25$

HOMEWORK 21D

1 **a** Draw the line $y = 3$ (as a solid line). **b** Shade the region defined by $y \geq 3$.

2 **a** Draw the line $x = -1$ (as a dashed line). **b** Shade the region defined by $x < -1$.

3 **a** Draw the line $x = -1$ (as a dashed line).
b Draw the line $x = 3$ (as a solid line) on the same grid.
c Shade the region defined by $-1 < x \leq 3$.

4 **a** On the same grid, draw the regions defined by the inequalities
i $-2 \leq x \leq 2$ **ii** $-1 < y \leq 3$
b Are the following points in the region defined by both inequalities?
i (2, 2) **ii** (–2, 2) **iii** (–2, –1) **iv** (–2, 3)

5 **a** Draw the line $y = 2x + 1$ (as a solid line).
b Shade the region defined by $y \leq 2x + 1$.

6 **a** Draw the line $3x + 4y = 12$ (as a dashed line).
b Shade the region defined by $3x + 4y > 12$.

7 **a** Draw the line $y = 2x - 1$ (as a solid line).
b Draw the line $x + y = 5$ (as a solid line) on the same diagram.
c Shade the diagram so that the region defined by $y \geq 2x - 1$ is left unshaded.
d Shade the diagram so that the region defined by $x + y \leq 5$ is left unshaded.
e Show clearly with an R the region defined by both inequalities.

8 **a** Shade $x \leq 2$, $y \geq x - 2$ and $x + y \geq -2$ on the same grid.
b Show clearly the region defined by all three inequalities by a letter R.

9 **a** On the same grid, draw the regions defined by the following inequalities. (Shade the diagram so that the overlapping region is left blank.)
i $y < x + 2$ **ii** $y \geq 2x - 2$ **iii** $y \geq 0$
b Are the following points in the region defined by all three inequalities?
i (0, 2) **ii** (0, –2) **iii** (2, 2) **iv** (4, 4)

★10 **a** On a pair of axes, leave unshaded the region represented by the following inequalities. **i** $x \leq 2$ **ii** $y > 1$ **iii** $y \leq x + 1$
b Write down the co-ordinates of all the points whose co-ordinates are integers and lie in the region which satisfy all the inequalities in part **a**.

HOMEWORK 21E

1 A local shop does A4 and A3 photocopies. A4 copies cost 6p. A3 copies cost 20p. How much will
a x A4 copies cost **b** y A3 copies cost
c x A4 and y A3 copies together cost?

2 A school canteen sells pizza at 50p a slice and chips at 75p per bag. How much do each of the following cost?
a 2 slices of pizza and a bag of chips. **b** x slices of pizza and y bags of chips.
c Frank has £5. How much change will he get if he buys x slices of pizza and 2 bags of chips? Can $x = 8$?

3 Apple computers make two main computers. The G3 and the G4. G3s cost £A and G4s cost £B. How much are
a x G3s and y G4s **b** x G3s and three times as many G4s
c x G3s and $(x + 5)$ G4s?

4 If $2x + y < 30$, which of the following may be true?
a $x + y = 10$ **b** $2x > 30$ **c** $2x + y \leq 40$

5 Tins of pears cost 60p. Tins of peaches cost 75p. Mary has £6.00. She buys x tins of pears and y tins of peaches. Explain why $4x + 5y < 40$.

6 Gwen is going on holiday for 14 days. She is taking x paperback novels and y magazines. It takes her 3 days to read a novel and 1 day to read a magazine. She gets bored with magazines so needs at least 3 novels. Because of space in her luggage she cannot take more than 7 books and magazines altogether.

a Explain why $3x + y \leq 14$, $x \geq 3$, $x + y \leq 7$

b Which of these combinations satisfies all the above requirements

i 3 books and 4 magazines **ii** 4 books and 3 magazines

iii 4 books and 4 magazines **iv** 2 books and 5 magazines?

★7 The school hall seats a maximum audience of 200 people for performances. Tickets for the school concert cost £2 or £3 each. The school needs to raise at least £450 from this concert. It is decided that the number of £3 tickets must not be greater than twice the number of £2 tickets. There are x tickets at £2 each and y tickets at £3 each.
Explain why **i** $x + y \leq 200$ **ii** $2x + 3y \geq 450$ **iii** $y \leq 2x$

HOMEWORK 21F

1 Computer Zip disks come in 2 sizes, 100 Mb and 250 Mb. 100 Mb disks cost £6 and 250 Mb cost £25 each. The company buys x 100 Mb disks and y 250 Mb disks. A company has a budget of £300 to spend on disks.

a Explain why $6x + 25y \leq 300$

b The company needs at least 4000 Mb of storage. Explain why $2x + 5y \geq 80$

c Which of these combinations of disks will satisfy the requirements

i (0, 16) **ii** (2, 15) **iii** (1, 16)?

2 Fred goes to the shop for some tins of peas and beans. Tins of beans cost 60p. Tins of peas cost 40p. He buys x tins of beans and y tins of peas.

a Fred has £3 to spend. Explain why $3x + 2y \leq 15$

b Fred's mum told him to make sure he has at least 2 more tins of peas than beans. Explain why $y - x \geq 2$

c Find a combination of tins that satisfy both requirements.

3 A library is restocking with books. Hardback books cost £10. Paperback books cost £8. They buy x hardback books and y paperback books.

a The library has £2000 to spend on books. Write down an inequality that satisfies this restriction. Simplify your answer.

b Hardback books take up twice as much space as paperbacks. The library only has space for the equivalent of 300 paperback books. Explain why $2x + y \leq 300$

c The library needs to have at least 175 new books. Write down an inequality that satisfies this requirement.

d Show that all these combinations satisfy all of the above requirements

i (115, 60) **ii** (0, 200) **iii** (64, 170) **iv** (0, 250)

e Which of the combinations in **d**

i takes up the most space **ii** is cheapest **iii** gives the most books?

4 Rachel keeps horses and donkeys. She has x donkeys and y horses.

a Donkeys cost £12 per week to keep and horses cost £15 per week to keep. Rachel has a weekly budget of £210 to feed her animals. Write down an inequality that satisfies this requirement. Simplify your answer as much as possible.

b Donkeys need at least 60 m^2 to graze and horses need at least 30 m^2 to graze. Rachel's field has an area of 600 m^2. Write down an inequality that satisfies this requirement.

c Rachel needs at least 2 horses. Find a combination of donkeys and horses that satisfies all of the above conditions.

5 Steve keeps finches and budgies. He has x budgies and y finches.

a Budgies cost 50p a week to feed and finches cost 60p a week to feed. Steve has £12 a week to spare for food. Write down and simplify an inequality that satisfies this restriction.

b Budgies need at least 0.2 cubic metres of space and finches need at least 0.15 cubic metres of space. Steve has a shed that has 4 cubic metres of space. Show that $4x + 3y \leq 80$

c To stop fighting there must be at least twice as many finches as budgies. Write down an inequality that satisfies this restriction.

d Can Steve keep

i 5 budgies and 16 finches **ii** 7 budgies and 13 finches

iii 6 budgies and 14 finches?

★6 A company is opening a site for caravans and tents. Each caravan requires $100\,m^2$ of land and each tent $50\,m^2$. There is a total of $2000\,m^2$ of land available. For each caravan rented, the firm makes £10 profit per night and for each tent £8 per night. The local council say that there can be no more than twice as many tents as caravans. x caravan plots and y tent plots are taken.

a Write down two inequalities which must be true.

b Find the maximum value of $x + y$, subject to all two conditions.

c How much profit does the company make?

HOMEWORK 21G

The situations in questions 1 to 5 below are identical to those in questions 1 to 5 in Homework 21F. Shade out the regions not required, leaving the feasible region blank. Label the feasible region as R.

1 Computer Zip disks come in 2 sizes, 100 Mb and 250 Mb. 100 Mb disks cost £6 and 250 Mb disks cost £25 each. The company buys x 100 Mb disks and y 250 Mb disks. The company has a budget of £300 to spend on disks. The company needs at least 4000 Mb of storage.

a Write down two inequalities from this information and represent them on the same grid.

b Which acceptable combination of disks gives the least number of disks in total?

c Which acceptable combination of disks gives the cheapest option?

d Which acceptable combination of disks gives the most storage?

2 Fred goes to the shop for some tins of peas and beans. Tins of beans cost 60p. Tins of peas cost 40p. Fred has £3.00 to spend. He buys x tins of beans and y tins of peas. Fred's mum told him to make sure he has at least 2 more tins of peas than beans.

a Find two inequalities that satisfy both conditions and represent them on the same grid.

b Fred's mum wants the cheapest option. Which acceptable combination of tins is this?

c Fred wants as many beans as possible. Which acceptable combination of tins is this?

3 A library is restocking with books. Hardback books cost £10. Paperback books cost £8. They buy x hardback books and y paperback books. The library has £2000 to spend on books. Hardback books take up twice as much space as paperbacks. The library only has space for the equivalent of 300 paperback books. The library needs to have at least 175 new books.

- **a** Find three inequalities that satisfy these requirements and represent them on the same grid.
- **b** Which of the acceptable combinations gives
  - **i** the most books **ii** the cheapest option **iii** the least amount of space used?

4 Rachel keeps horses and donkeys. She has x donkeys and y horses. Donkeys cost £12 per week to keep and horses cost £15 per week to keep. Rachel has a weekly budget of £210 to feed her animals. Donkeys need at least $60\,m^2$ to graze and horses need at least $30\,m^2$ to graze. Rachel's field has an area of $600\,m^2$. Rachel needs at least two horses.

- **a** Find three inequalities that satisfy these requirements and represent them on the same grid.
- **b** Which acceptable combination gives the most animals?
- **c** Which acceptable combination gives the cheapest option?

5 Steve keeps finches and budgies. He has x budgies and y finches. Budgies cost 50p a week to feed and finches cost 60p a week to feed. Steve has £12 a week to spare for food. Budgies need at least 0.2 cubic metres of space and finches need at least 0.15 cubic metres of space. Steve has a shed that has 4 cubic metres of space. To stop fighting there must be at least twice as many finches as budgies.

- **a** Find three inequalities that satisfy these requirements and represent them on the same grid.
- **b** Which acceptable combination gives the most birds?
- **c** Which acceptable combination gives the cheapest option?

6 A local garage does a car cleaning service. They offer standard and deluxe services. The standard costs £50 and takes 4 hours. The deluxe costs £90 and takes 8 hours. The garage is open for 60 hours a week. The garage has a contract with a local taxi firm to do at least 4 standard and 2 deluxe cleans a week. In a week the garage does x standard and y deluxe cleans.

- **a** Explain why $x + 2y \leq 15$
- **b** Write down two other inequalities that must be true.
- **c** Show all the inequalities on the same grid.
- **d** Find the combination of cleaning services that makes the most money.

7 A chef takes 1 hour to prepare a batch of chips and 2 hours to prepare a batch of fish. He works 10 hours each day. The kitchen can only manage a total of 8 batches of food per day. The chef has a deal with a local factory to supply 4 batches of chips and 2 batches of fish per day. The chef makes £25 profit on a batch of chips and £40 profit on a batch of fish. By drawing suitable inequalities on a graph find the combination that meets all the above requirements and gives the chef the maximum profit.

★8 A council decides to buy some boats for use on their boating lake. They decide to buy at least 4, but not more than 8 canoes and at least 3 dinghies. The maximum number of boats allowed on the lake is 14.

- **a** Using c as the number of canoes and d as the number of dinghies, write down four inequalities which represent the conditions given above.
- **b** Illustrate these four regions on the same grid in order to show, by not shading, the region which gives the possible combinations of canoes and dinghies.

c The council are prepared to spend a maximum of £1000 on these boats. Each canoe will cost £100 and each dinghy £80. Use the graphs to find the largest total number of boats they can buy.

Chapter 22 Vectors

HOMEWORK 22A

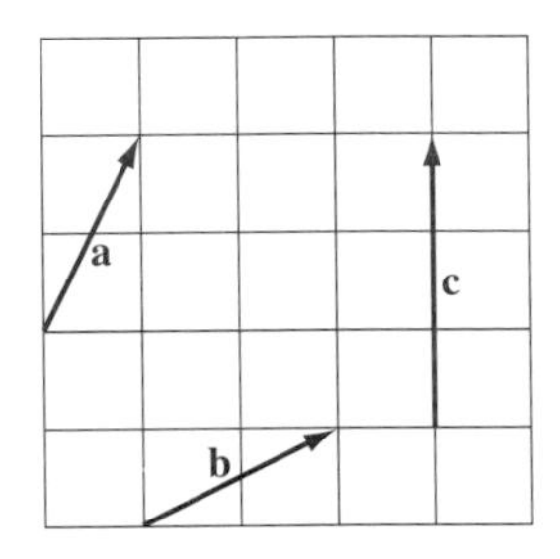

1 Vectors a, **b** and **c** are shown on the diagram.

a Draw the vectors represented by

i $\mathbf{a} + \mathbf{b}$ **ii** $-\mathbf{a}$ **iii** $\mathbf{a} - \mathbf{b}$ **iv** $\mathbf{b} - \mathbf{a}$ **v** $-\mathbf{b}$

vi $-\mathbf{a} - \mathbf{b}$

b Explain the connection between the answers to parts **i** and **vi** and parts **iii** and **iv**

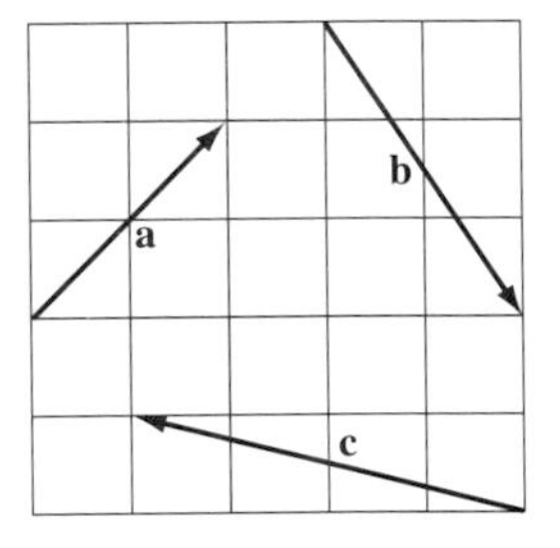

2 The diagram shows vectors **a**, **b** and **c**.

a Draw the vectors shown by **i** $\mathbf{a} + \mathbf{b}$ **ii** $\mathbf{a} + \mathbf{b} + \mathbf{c}$

b Explain the answer to **a** part **ii**

c Write the vectors **a**, **b** and **c** as column vectors and then write as column vectors

i $\mathbf{a} - \mathbf{b}$ **ii** $\mathbf{a} - \mathbf{c}$ **iii** $\mathbf{a} + \mathbf{b} + \mathbf{c}$

3 The diagram shows three vectors **a**, **b** and **c**.

a is $\overrightarrow{OA}$, **b** is $\overrightarrow{OB}$ and **c** is $\overrightarrow{OC}$.

a is the vector $\binom{2}{2}$, **b** is the vector $\binom{4}{1}$

c is the vector $\binom{6}{0}$

a Write as a column vector and show on a diagram the vectors

i $\overrightarrow{AB}$ **ii** $\overrightarrow{AC}$ **iii** $\overrightarrow{BC}$

b What do your answers to **a** tell you?

c Would parts **i** and **ii** of **a** be enough to tell you that ABC were on a straight line? Explain your answer.

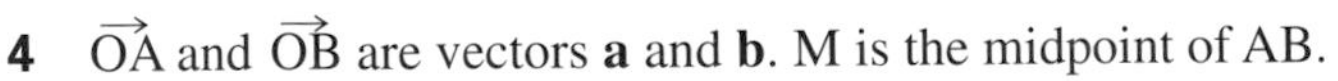

4 $\overrightarrow{OA}$ and $\overrightarrow{OB}$ are vectors **a** and **b**. M is the midpoint of AB.

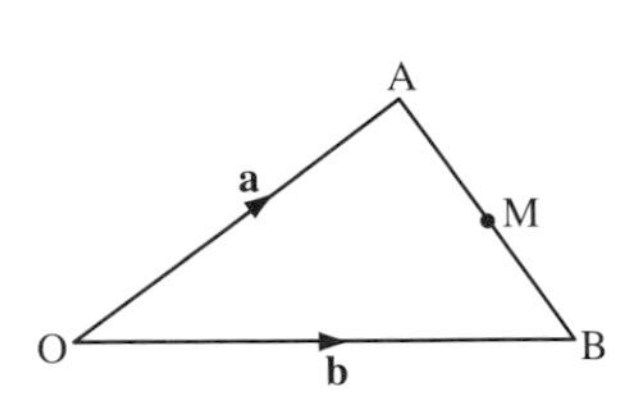

a Express in terms of **a** and **b** the vectors

i $\overrightarrow{AB}$ **ii** $\overrightarrow{AM}$ **iii** $\overrightarrow{OM}$

b Draw on a copy of the diagram the points X and Y such that $\overrightarrow{OX} = 2\mathbf{a} + \mathbf{b}$ and $\overrightarrow{OY} = \mathbf{a} + 2\mathbf{b}$

Express $\overrightarrow{XY}$ in terms of **a** and **b**.

c What other vector on the diagram is equivalent to $\overrightarrow{XY}$?

5 The diagram shows vectors **a**, **b**, **c**, **d**, **e**, **f**, **g** and **h**.

a Find a pair of vectors that are equal. Write them as column vectors.

b Find two pairs of vectors that are opposite in direction and the same size. Write them as column vectors. Describe how you could tell from column vectors that two vectors are opposite and of the same length.

c Find two pairs of vectors that are parallel where one is twice the length of the other. Write them as column vectors. Describe how you can tell from column vectors if two vectors are parallel and one is twice the length of the other.

d Find two pairs of vectors that are the same length but perpendicular. Write them as column vectors. Describe how you can tell from the column vectors that two vectors are perpendicular and the same length.

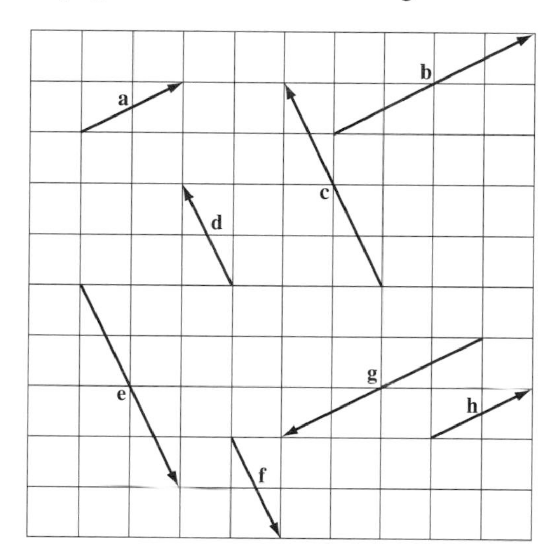

6 $\overrightarrow{OA}$ and $\overrightarrow{OB}$ are the vectors **a** and **b**. R is the midpoint of AB. P divides OB in the ratio 3:1 and Q divides OA in the ratio 2:1. Express in terms of **a** and **b**.

a $\overrightarrow{OP}$ **b** $\overrightarrow{OQ}$ **c** $\overrightarrow{OR}$ **d** $\overrightarrow{RP}$ **e** $\overrightarrow{QR}$ **f** $\overrightarrow{PQ}$

g Add $\overrightarrow{QR} + \overrightarrow{RP} + \overrightarrow{PQ}$. Explain your result.

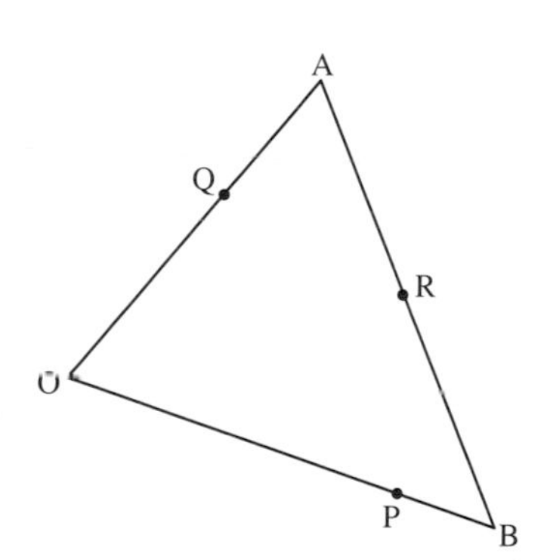

7 OACB is a trapezium, where $\overrightarrow{OA} = \mathbf{a}$, $\overrightarrow{OB} = \mathbf{b}$ and $\overrightarrow{BC} = 2\mathbf{a}$. P and Q are the midpoints of $\overrightarrow{OB}$ and $\overrightarrow{AC}$. Express in terms of **a** and **b**.

a $\overrightarrow{OP}$ **b** $\overrightarrow{AQ}$ **c** $\overrightarrow{PQ}$ **d** How can you tell that $\overrightarrow{PQ}$ is parallel to $\overrightarrow{OA}$?

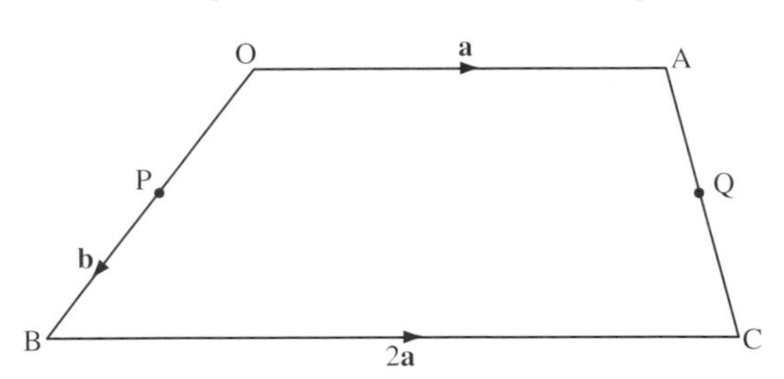

8 $\overrightarrow{OA}$, $\overrightarrow{OB}$ are the vectors **a** and **b**. C is the point on AB such that C is $\frac{3}{4}$ along AB.

a Express $\overrightarrow{OC}$ in terms of **a** and **b**.

b If D is the point that is $\frac{2}{3}$ along AC, write down the vector $\overrightarrow{OD}$.

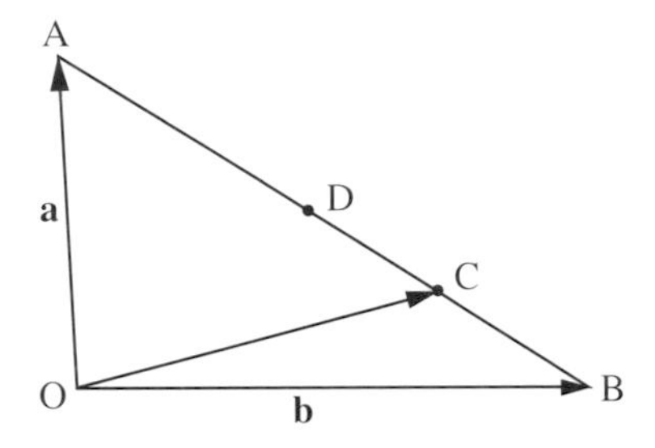

★9 In the diagram, X is the point on AB such that AX = 4XB. Given that $\overrightarrow{OA} = 10\mathbf{q}$ and $\overrightarrow{OB} = 5\mathbf{p}$, express in terms of **p** and/or **q**,

a $\overrightarrow{AB}$ **b** $\overrightarrow{AX}$ **c** $\overrightarrow{OX}$

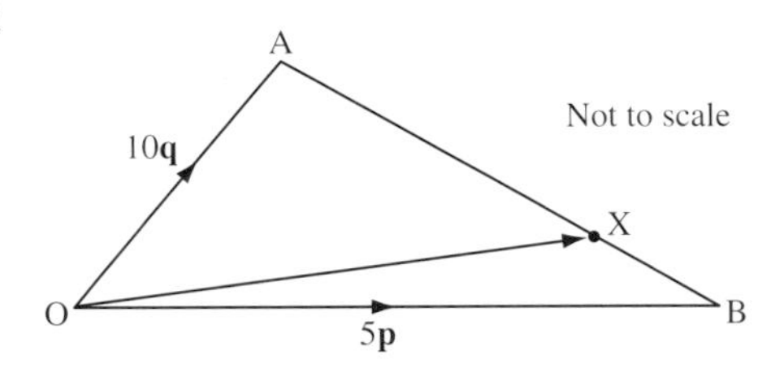

HOMEWORK 22B

1 OACB is a rectangle. $\overrightarrow{OA} = \mathbf{a}$ and $\overrightarrow{OB} = \mathbf{b}$. Q is the midpoint of BC and P divides BA in the ratio 1 : 2. Find the vectors

a $\overrightarrow{BP}$ **b** $\overrightarrow{OP}$ **c** $\overrightarrow{OQ}$

d Explain the relationship between O, P and Q.

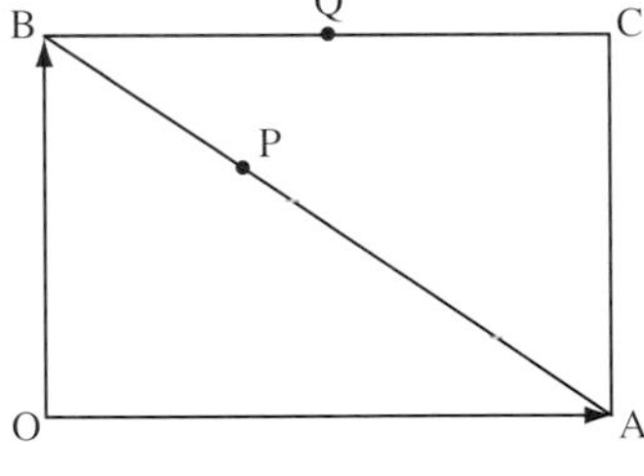

2 $\overrightarrow{OA} = \mathbf{a}$. $\overrightarrow{OB} = \mathbf{b}$. P is the point that divides OB in the ratio 1 : 2. Q is the point that divides OA in the ratio 2 : 1.

a Express in terms of **a** and **b** **i** $\overrightarrow{AP}$ **ii** $\overrightarrow{BQ}$

b Explain why OR can be written as $\mathbf{a} + n\overrightarrow{AP}$.

c Explain why OR can be written as $\mathbf{b} + m\overrightarrow{BQ}$.

d Show that the expressions in parts **b** and **c** are equal when $n = \frac{3}{7}$ and $m = \frac{6}{7}$.

e Hence find the vector $\overrightarrow{OR}$ in terms of **a** and **b**.

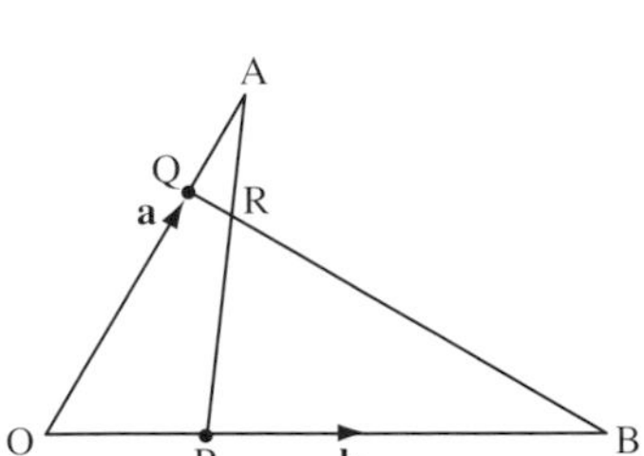

3 OAB is a triangle. $\overrightarrow{OA} = \mathbf{a}$. $\overrightarrow{OB} = \mathbf{b}$. R is the midpoint of AB. Q is the midpoint of OA. Find in terms of **a** and **b** the vectors

a $\overrightarrow{OR}$ **b** $\overrightarrow{QB}$

c G is the point where OR and QB meet. Explain why $\overrightarrow{OG}$ can be written both as $n\overrightarrow{OR}$ and $\frac{1}{2}\mathbf{a} + m(\overrightarrow{QB})$.

d You are given that $m = n$.
Find values of m and n that satisfy the equations in **c**.

e Hence express $\overrightarrow{OG}$ in terms of **a** and **b**.

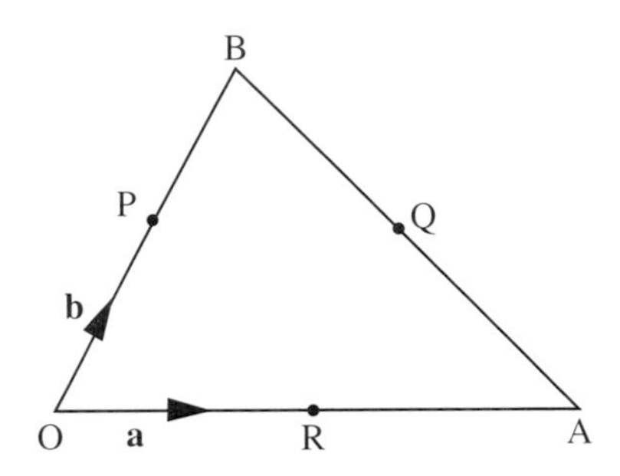

4 OAB is a triangle. P, Q and R are the midpoints of OB, OA and AB. $\overrightarrow{OR} = \mathbf{a}$ and $\overrightarrow{OP} = \mathbf{b}$.

a Express in terms of **a** and **b** the vectors
i $\overrightarrow{RP}$ **ii** $\overrightarrow{AB}$

b What do your answers to part **a** tell you about the relationship between the lines RP and AB?

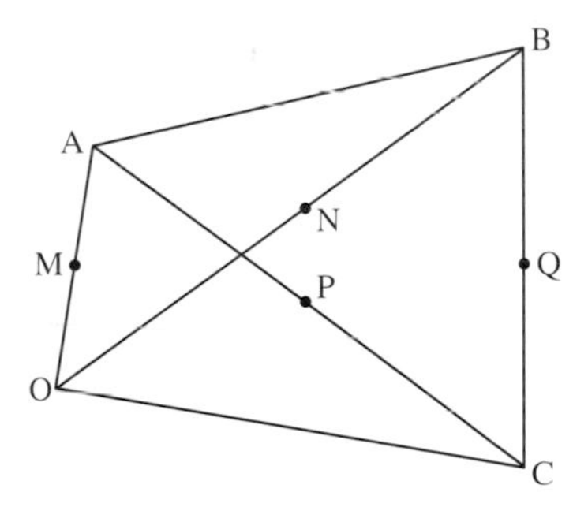

5 OAC is a triangle such that $\overrightarrow{OC} = 12\mathbf{q}$, $\overrightarrow{OB} = 3\mathbf{q}$ and $\overrightarrow{OA} = 3\mathbf{p}$.

a Find in terms of **p** and **q** the vectors
i AB **ii** AC

b Given that $AQ = \frac{1}{3}AC$ express $\overrightarrow{OQ}$ in terms of **p** and **q**.

c Given that $\overrightarrow{OR} = \mathbf{p} + 2\mathbf{q}$ what can you say about the points O, R and Q?

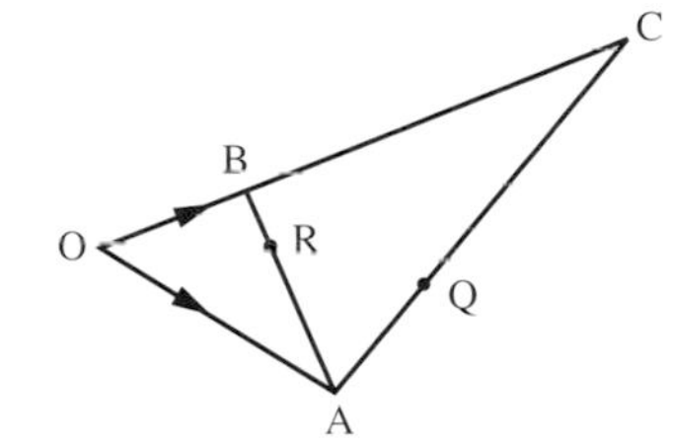

6 OABC is a quadrilateral. $\overrightarrow{OA} = \mathbf{a}$, $\overrightarrow{OB} = \mathbf{b}$, $\overrightarrow{OC} = \mathbf{c}$. M, N, Q and P are the midpoints of OA, OB, CB and OC.

a Find in terms of **a**, **b** and **c** the vectors
i $\overrightarrow{BC}$ **ii** NQ **iii** $\overrightarrow{MP}$

b What type of quadrilateral is MNQP? Explain your answer.

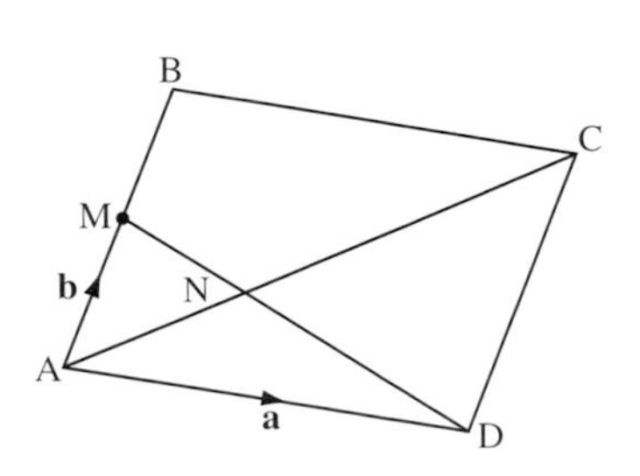

7 ABCD is a parallelogram. $\overrightarrow{AD} = \mathbf{a}$ and $\overrightarrow{AB} = \mathbf{b}$. M is the midpoint of AB and N is the point where MD meets AC.

a Find in terms of **a** and **b** the vectors
i $\overrightarrow{MC}$ **ii** $\overrightarrow{MD}$ **iii** $\overrightarrow{AC}$

b Set up a pair of simultaneous equations for AN (Hint use $n\overrightarrow{AC}$ and $\overrightarrow{AM} + m\overrightarrow{MD}$) to find the vector $\overrightarrow{AN}$.

c Find **i** $\overrightarrow{CN} + \overrightarrow{DN}$ **ii** $\overrightarrow{AN} + \overrightarrow{MN}$

d What conclusion can you draw about the answers to part **c**?

B
C
M
b
N
A
a
D

★**8** OACB and OBRS are parallelograms.
$\overrightarrow{OA}$ is **a**, $\overrightarrow{OB}$ is **b**, and $\overrightarrow{BR}$ is **r**. Find in terms of **a**, **b**, and **r** expressions in their simplest forms for

a $\overrightarrow{OR}$

b $\overrightarrow{SB}$

c $\overrightarrow{OX}$, where X is the mid-point of AR, i.e. $\frac{1}{2}(\overrightarrow{OA} + \overrightarrow{OR})$

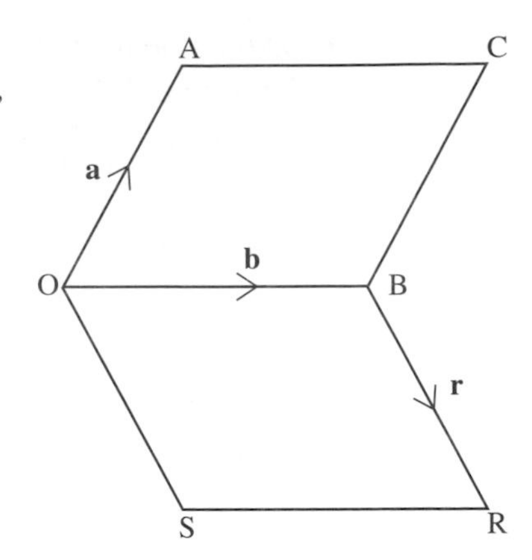

Chapter 23 Transformation of graphs

HOMEWORK 23A

Do not use a graphical calculator.

1 On the same axes sketch the graphs of

a $y = x^2$ **b** $y = 2x^2$ **c** $y = x^2 + 2$ **d** $(x + 2)^2$

e Describe the transformation(s) that take(s) the graph in part **a** to each of the graphs in parts **b** to **d**.

2 On the same axes sketch the graphs of

a $y = x^2$ **b** $y = 3x^2 + 2$ **c** $y = x^2 - 3$ **d** $y = \frac{1}{2}x^2 + 1$

e Describe the transformation(s) that take(s) the graph in part **a** to each of the graphs in parts **b** to **d**.

3 On the same axes sketch the graphs of

a $y = x^2$ **b** $y = (x + 4)^2$ **c** $y = -x^2$ **d** $y = 2 - x^2$

e Describe the transformation(s) that take(s) the graph in part **a** to each of the graphs in parts **b** to **d**.

4 On the same axes sketch the graphs of

a $y = \sin x$ **b** $y = 3 \sin x$ **c** $y = \sin x + 3$ **d** $y = \sin (x + 30)$

e Describe the transformation(s) that take(s) the graph in part **a** to each of the graphs in parts **b** to **d**.

5 On the same axes sketch the graphs of

a $y = \sin x$ **b** $y = - \sin x$ **c** $y = \sin \frac{x}{3}$ **d** $y = 3 \sin \frac{x}{2}$

e Describe the transformation(s) that take(s) the graph in part a to each of the graphs in parts **b** to **d**.

6 On the same axes sketch the graphs of

a $y = \sin x$ **b** $y = 3 \sin x$ **c** $y = \sin (x + 45°)$ **d** $y = 2 \sin (x + 90°)$

e Describe the transformation(s) that take(s) the graph in part **a** to the graphs in parts **b** to **d**.

7 On the same axes sketch the graphs of

a $y = \cos x$ **b** $y = -\cos x$ **c** $y = \cos x + 4$ **d** $y = 2\cos x$

e Describe the transformation(s) that take(s) the graph in part **a** to each of the graphs in parts **b** to **d**.

8 On the same axes sketch the graphs of

a $y = \cos x$ **b** $y = 3\cos x$ **c** $y = \cos(x + 60°)$ **d** $y = 2\cos x + 3$

e Describe the transformation(s) that take(s) the graph in part **a** to each of the graphs in parts **b** to **d**.

9 Explain why the graphs of $y = \cos x$ and $y = \sin(x + 90°)$ are the same.

★10 The table shows some values of the function $f(x) = (x - 2)^2 + 4$, where $-3 < x < 4$.

a Draw the graph of $y = f(x)$.

b On the same axes, draw the graph of $y = x^2$.

x	−3	−2	−1	0	1	2	3	4
f(x)	29	20	13	8	5	4	5	8

c Describe how the graph of $y = (x - 2)^2 + 4$ can be obtained from the graph $y = x^2$ by a transformation. State clearly what this transformation is.

HOMEWORK 23B

Do not use a graphical calculator.

1 What is the equation of the graph obtained when the following transformations are performed on the graph of $y = x^2$?

a Stretch by a factor of 3 in the y-direction.

b Translation of $\begin{pmatrix}0\\2\end{pmatrix}$.

c Translation of $\begin{pmatrix}-4\\0\end{pmatrix}$.

d Stretch by a factor of 2 in the y-direction followed by a translation of $\begin{pmatrix}0\\3\end{pmatrix}$.

e Translation of $\begin{pmatrix}3\\-2\end{pmatrix}$.

f Stretch, scale factor 2 in the y-direction, followed by a reflection in the x-axis.

2 What is the equation of the graph obtained when the following transformations are performed on the graph of $y = \sin x$?

a Stretch by a factor of 2 in the y-direction.

b Translation of $\begin{pmatrix}0\\2\end{pmatrix}$.

c Translation of $\begin{pmatrix}60\\2\end{pmatrix}$.

d Stretch by a factor of $\frac{1}{2}$ in the y-direction followed by a translation of $\begin{pmatrix}0\\-1\end{pmatrix}$.

e Translation of $\begin{pmatrix}90\\-1\end{pmatrix}$.

3 The graphs below are all transformations of $y = x^2$. Two points through which each graph passes are indicated. Use this information to work out the equation of each graph.

a

y
(0, 3)
(1, 4)
O
x

b

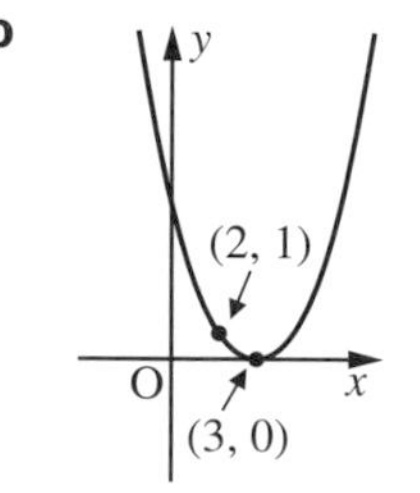

c

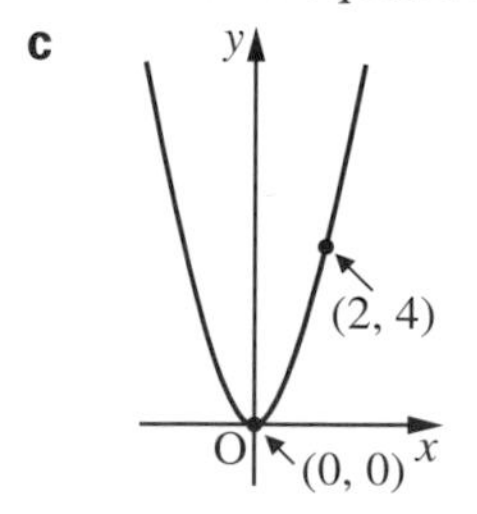

d

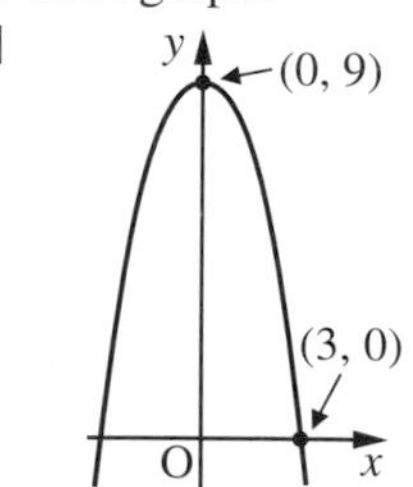

4 The graphs below are all transformations of $y = \sin x$. Two points through which each graph passes are indicated. Use this information to work out the equation of each graph.

a

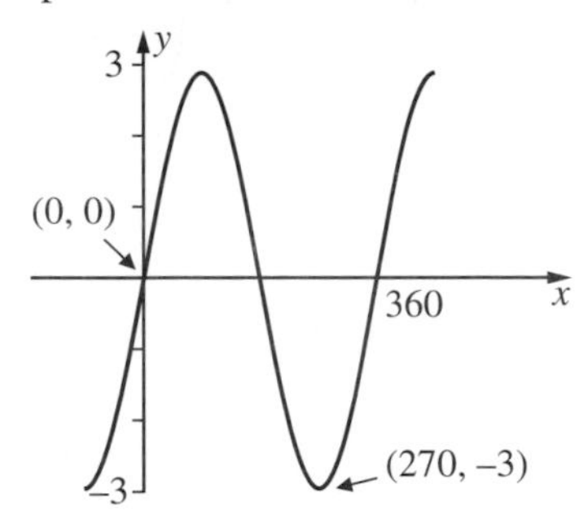

b

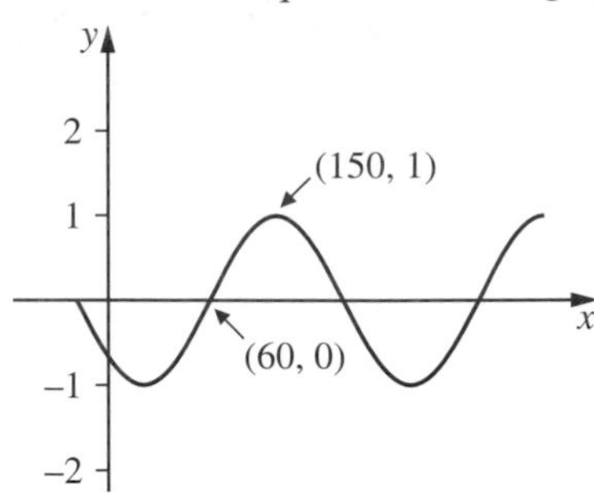

c

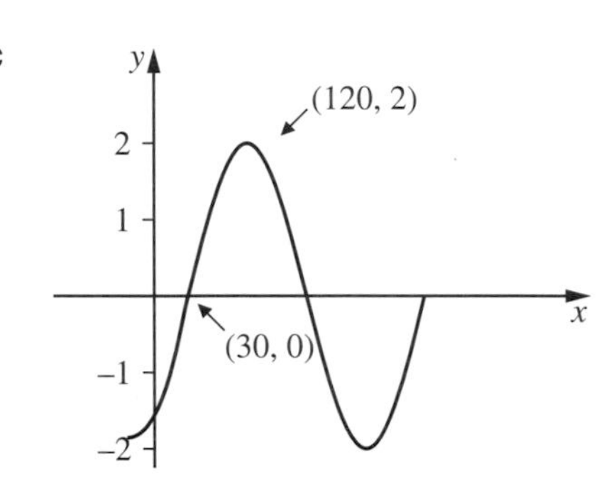

d

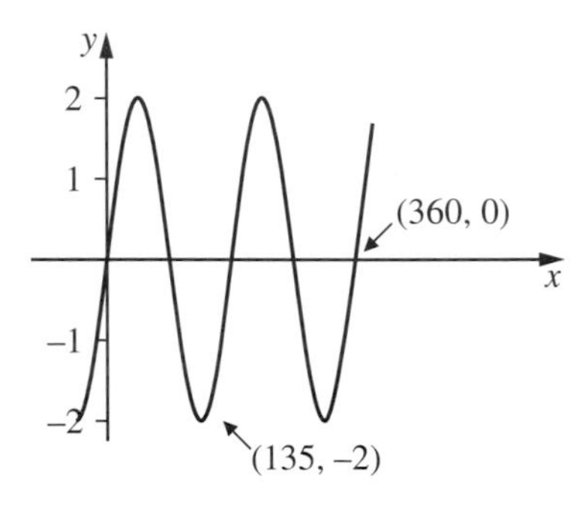

5 Below are the graphs of $y = -\sin x$ and $y = \cos x$.

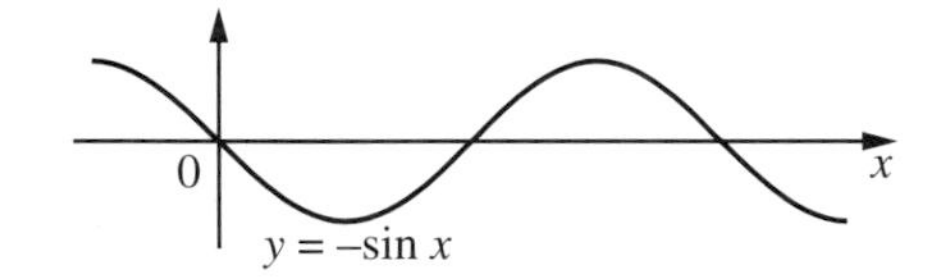

$y = \cos x$

a Describe a series of transformations that would take the first graph to the second.

b Which of the following is equivalent to $y = -\sin x$?

i $y = \sin (x + 180°)$ **ii** $y = \cos (x + 90°)$ **iii** $y = 2 \sin \frac{x}{2}$

6 Match each of the graphs A, B ,C, D and E to one of these equations.

i $y = 2x^2$ **ii** $y = -x^2 + 4$ **iii** $y = -(x + 2)^2$ **iv** $y = (x - 2)^2$ **v** $y = x^2 + 2$

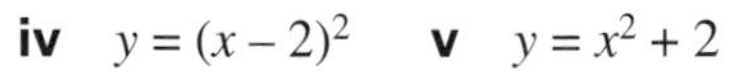

a

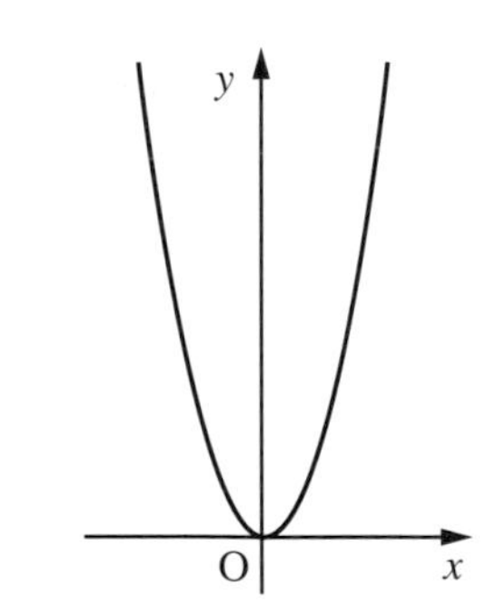

b

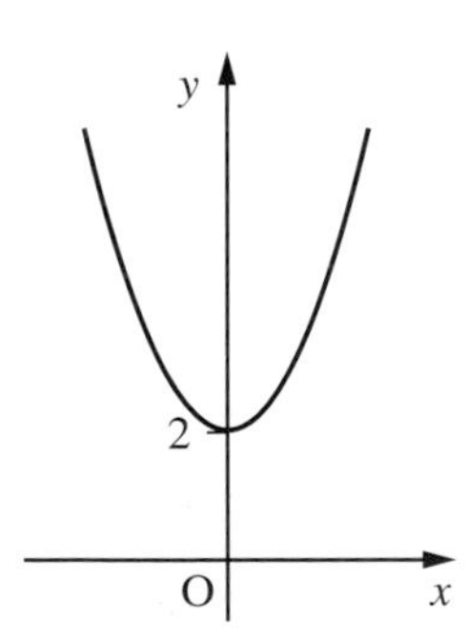

c

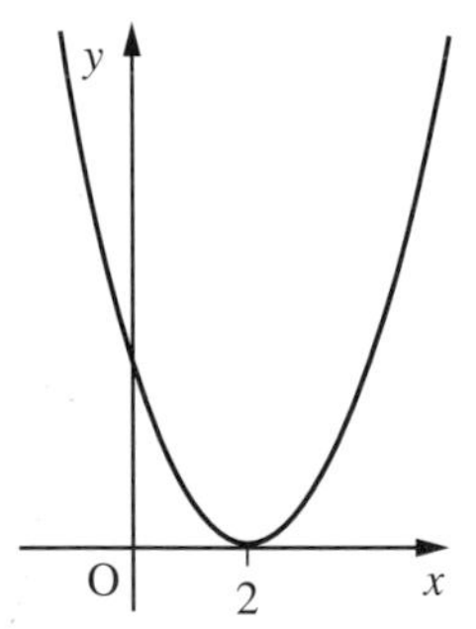

d

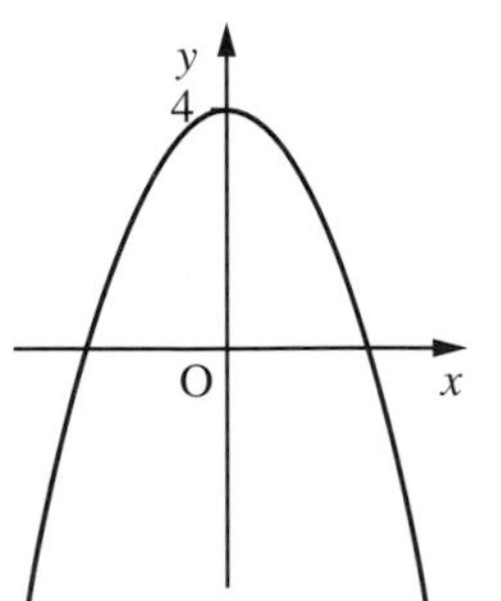

e

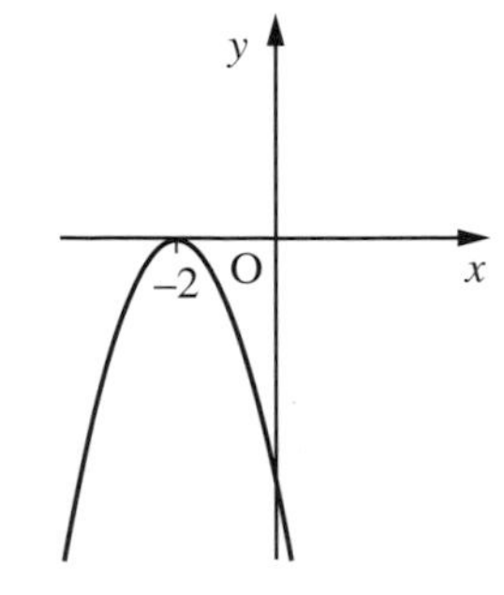

★**7** The graph of $y = f(x)$ has been drawn.
Sketch the graph of
a $y = f(x) - 2$ **b** $y = f(x - 2)$

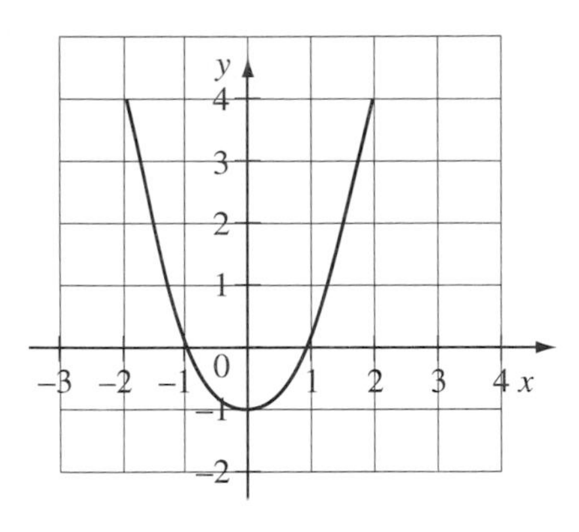

Chapter 24 Proof

Only one set of questions is given for this chapter. Questions should be chosen that are appropriate to current work.

1 Hollow squares are surrounded by 1 centimetre cubes.
e.g. this is a 5×5 hollow square.
Prove that when a $m \times m$ hollow square is placed inside a $n \times n$ hollow square the amount of space left is $n^2 - 4(n + m) + 8$

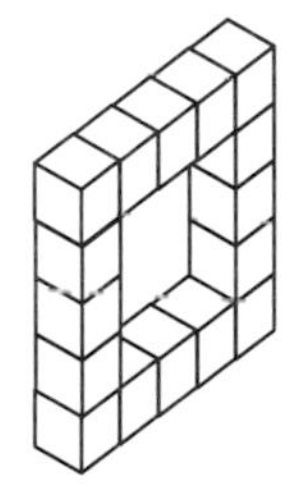

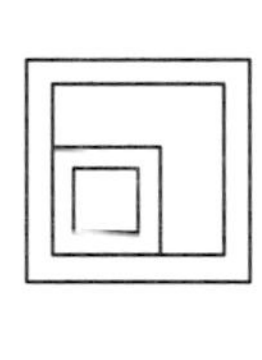

2 If m and n are integers then $(m^2 - n^2)$, $2mn$ and $(m^2 + n^2)$ will form three sides of a right angled triangle. e.g. Let $m = 5$ and $n = 3$, $m^2 - n^2 = 16$, $2mn = 30$, $m^2 + n^2 = 34$ and $34^2 = 1156$, $16^2 + 30^2 = 256 + 900 = 1156$
Prove this result.

3 Explain why the triangle number sequence 1, 3, 6, 10, 15, 21, 28, ...
follows the pattern of two odd numbers followed by two even numbers.

4 $10p + q$ is a multiple of 7. Prove that $3p + q$ is also a multiple of 7.

5 ABCD is a trapezium. The diagonals DB and AC meet at E.
Prove that the triangles ADE and BCE are equal in area.

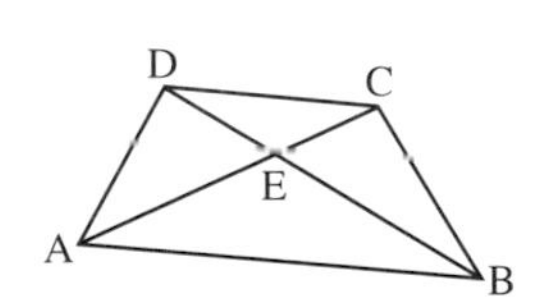

6 **a** Show that $2(5(x - 2) + y) = 10(x - 1) + 2y - 10$
b Prove that this trick works:
Think of two numbers less than 10.
Subtract 2 from the larger number and then multiply by 5.
Add the smaller number and multiply by 2.
Add 9 and subtract the smaller number.
Add 1 to both the tens digits and the units digits to obtain the numbers first thought of.

c Prove why the following trick works.

Choose two numbers. One with one digit the other with 2 digits.
Subtract 9 times the first number from 10 times the second number.
The units digit of the answer is the single digit number chosen and the sum of the other digits plus the units digit is the other number chosen.
e.g. Choose 7 and 23. $(23 \times 10) - (7 \times 9) = 167$.
The single digit number chosen is 7 the two digits number chosen is $16 + 7 = 23$.

7 ABCD is a rectangle. CEF is a triangle congruent to triangle ACD. BCE is a straight line. The line AC is extended to meet EF at P. Prove that AP is perpendicular to EF.

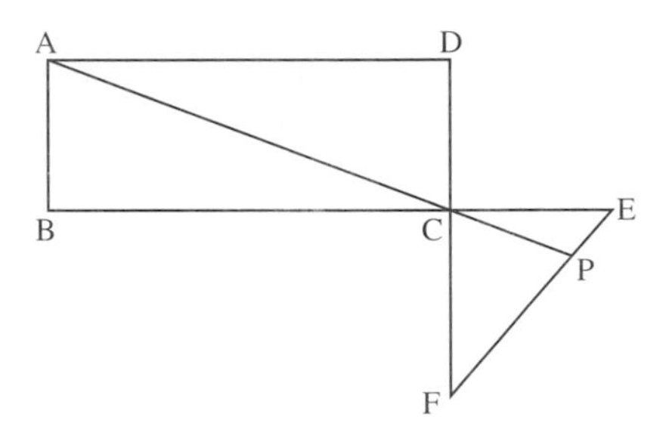

8 Prove that $(3n - 1)^2 + (3n)^2 + (3n + 1)^2 = (5n)^2 + (n - 1)^2 + (n + 1)^2$

9 Prove that $(n - 1)^2 + n^2 + (n + 1)^2 = 3n^2 + 2$

10 **a** What is the mth term of the sequence 4, 9, 14, 19, 24, … ?
b What is the mth term of the sequence 5, 10, 15, 20, 25, … ?
c If T_n represent the nth triangle number prove that $T_n = \frac{1}{2}n(n+1)$
d Prove that T_n is a multiple of 5 when n is a member of the series
4, 5, 14, 15, 19, 20, 24, 25, …

11 If T_n is any triangle number prove that
$3T_n = T_{2n+1} - T_{n+1}$ e.g. $T_4 = 10, T_9 = 45, T_5 = 15; 3 \times 10 = 45 - 15$

12 An isosceles trapezium is cut along a diagonal and the pieces put together as shown.

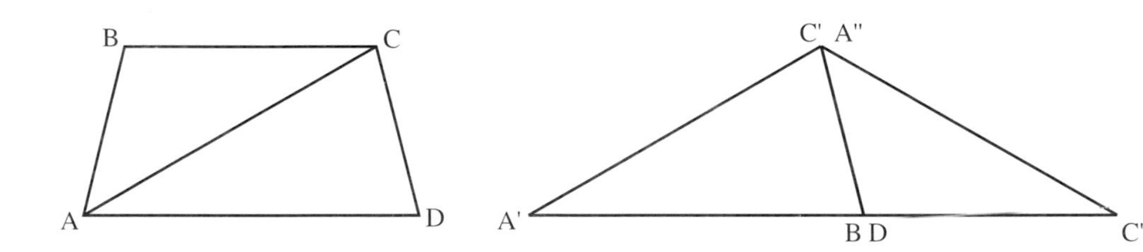

a Prove that A′BC″ is a straight line. **b** Prove that the triangle is isosceles.
c Explain why the triangle formed could never be equilateral.

13 If T_n is any triangle number prove that $\frac{T_n - 1}{(T_n)} = \frac{(n - 1)(n + 2)}{n(n + 1)}$

14 **a** What is the nth term of the sequence 1, 4, 7, 10, 13, 16, … .
b Explain why there is no multiple of 3 in the sequence.
c Prove that the sum of any 5 consecutive numbers in the sequence is a multiple of 5.

15 This question was first set in an examination in 1929.

$10^x = \frac{a}{b}$, $10^y = \frac{b}{a}$, Prove that $x + y = 0$

16 **a** Prove the alternate segment theorem.

b Two circles touch internally at T. The common tangent at T is drawn. Two lines TAB and TXY are drawn from T. Prove that AX is parallel to BY.

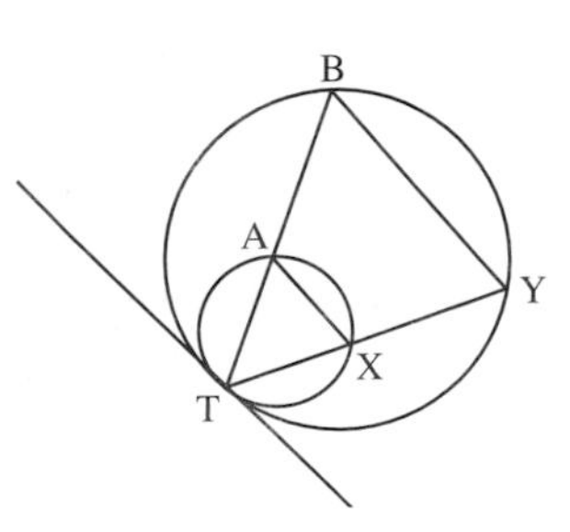

17 Two circles touch externally at T. A line ATB is drawn through T. The common tangent at T and the tangents at A and B meet at P and Q. prove that PB is parallel to AQ.

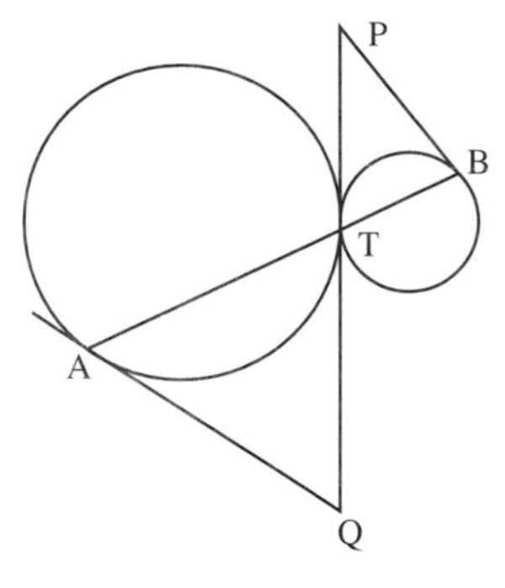

18 m and n are integers. Explain why

a the product $n(n + 1)$ must be even **b** $2m + 1$ is always an odd number.

c Look at the following numbers pattern

$1^2 - 1 = \mathbf{0}$
$2^2 - 1 = 3$
$3^2 - 1 = \mathbf{8}$
$4^2 - 1 = 15$
$5^2 - 1 = \mathbf{24}$
$6^2 - 1 = 35$

i Extend the pattern for 5 more lines to show that alternate values are multiples of 8.

ii Prove that this is true.

19 The midpoints of the edges of a square are joined to a vertex to create a smaller square (shown shaded). Explain why the shaded square has an area one-fifth of the area of the larger square.

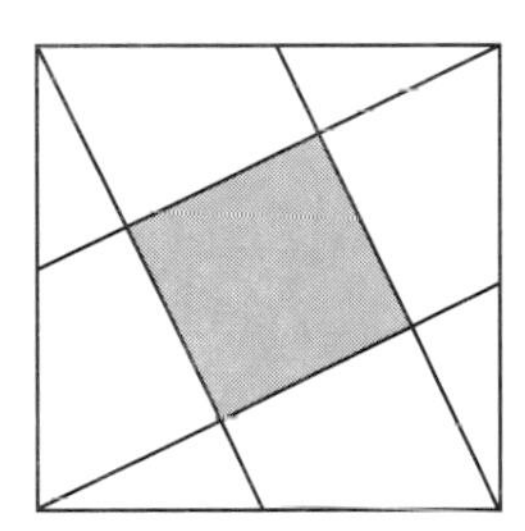

20 ABC is a triangle. P is a point on BC such that angle APC = angle BAC. The sides of ABC are a, b and c. AP = p and PC = r.

Prove that $r = \dfrac{b^2}{a}$.

(Hint: Show that triangles ABC and APC are similar).

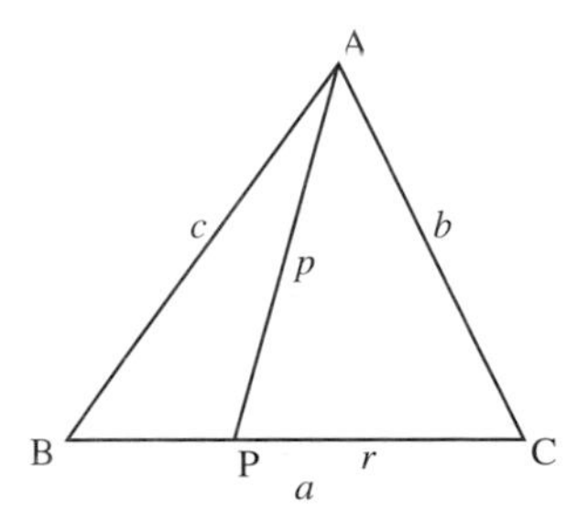

21 a, b, c and d are four consecutive integers. Prove that

a $bc - ad = 2$ **b** $ab + bc + cd + da + 1$ is a square number.

22 Prove that the square of the sum of two consecutive integers minus the sum of the squares of the two integers is four times a triangular number.
e.g. Let the two integers be 6 and 7. $(6 + 7)^2 - (6^2 + 7^2) = 169 - 85 = 84 = 4 \times 21$

23 Take any 4 numbers from the Fibonacci sequence, for example 8, 13, 21, 34
Prove that the following results are true.

a The sum of the first and fourth terms is double the third.
e.g. $34 + 8 = 42 = 2 \times 21$

b The difference between the first and fourth terms is double the second term.
e.g. $34 - 8 = 26 = 2 \times 13$

c The difference between the squares of the second and third terms is equal to the product of the first and fourth terms.
e.g. $21^2 - 13^2 = 272 = 8 \times 34$

24 a, b, c, d are consecutive integers. Prove that $bd - ac$ is always odd.

25 **a** Prove that the angles subtended by the chord at the circumference of a circle are equal.

b PQRS is a cyclic quadrilateral. PR and QS meet at T. Angles x, $2x$, $3x$ and $5x$ are marked on the diagram.

i Find x.

ii Show that the angles of the quadrilateral and angle STP form a number sequence.

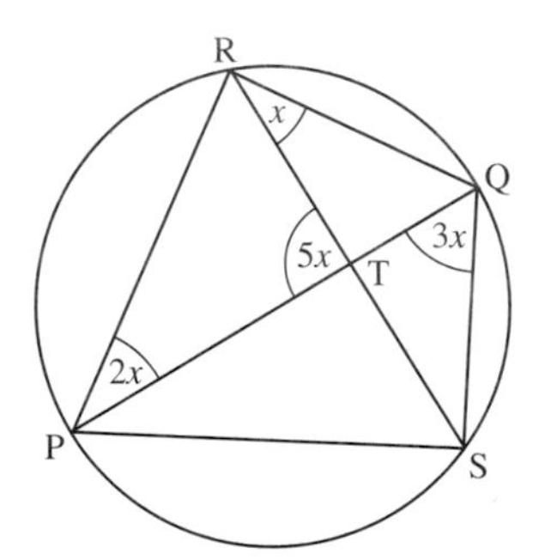

26 p, q and r are three consecutive numbers. Prove that $pr = q^2 - 1$.

27 ACB and ADB are right angled triangles. The lengths are as marked.

a Use Pythagoras' theorem to show that $x^2 = r^2 + s^2$

b Use Pythagoras' theorem on both triangles ACB and ADB to prove that $xt = sy$.

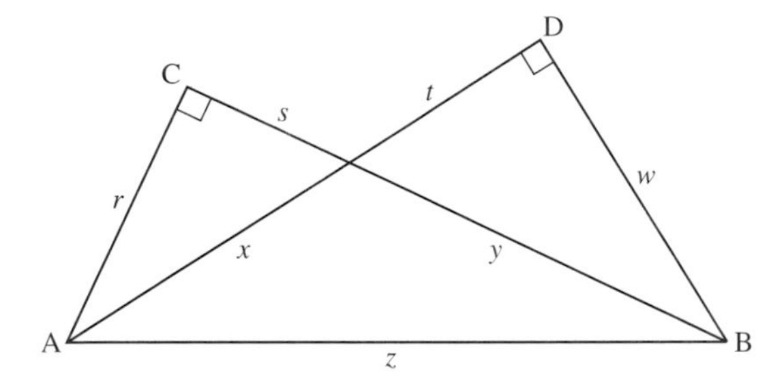